FLORIDA STATE
UNIVERSITY LIBRARIES

MAR 19 1998

TALLAHASSEE, FLORIDA

# POTIONS

# POISONS

# AND PANACEAS

*An Ethnobotanical Study of Montserrat*

David Eric Brussell

Foreword by Richard Evans Schultes
Preface by J.-P. Theurillat

*Southern Illinois University Press*
Carbondale and Edwardsville

To Prem Rawat
and Esther Trout

Copyright © 1997 David Eric Brussell
All rights reserved
Printed in the United States of America
00 99 98 97    4 3 2 1

All photographs not indicated otherwise by David Eric Brussell

"Mystery—South of Us" from *Floridays,* by Don Blanding (Port Salerno, Florida: Valentine Books, 1977). Copyright 1969 by Security-First National Bank. Reprinted with permission.

Library of Congress Cataloging-in-Publication Data

Brussell, David Eric
   Potions, poisons, and panaceas : an ethnobotanical study of Montserrat / David Eric Brussell : foreword by Richard Evans Schultes; preface by J.-P. Theurillat.
      p.   cm.
   Includes bibliographical references and indexes.
   1. Ethnobotany—Montserrat.  2. Botany—Montserrat.  3. Plants, Useful—Montserrat.  4. Folk medicine—Montserrat.  5. Montserrat—Social life and customs     I. Title.
GN560.M66B78   1997
581.6′1′0972975—dc20           90-25873
ISBN 0-8093-1552-1                    CIP
ISBN 0-8093-1553-X (pbk.)

The paper used in this publication meets the minimum requirements of American National Standard for Information Sciences—Permanence of Paper for Printed Library Materials, ANSI Z39.48-1984. ∞

## MYSTERY—SOUTH OF US

Florida thrusts like a guiding thumb
To the southern islands of rumba and rum,
To the mystery-cities and haunted seas
Of the Spanish Main and the Caribbees.
Where the ghosts of Columbus and
    Pirates Bold
Seek the Islands of Spice and the Streets
    of Gold,
Where the wandering phantom of Ku-Kul-Kan
Haunts the temples he builded in Yucatan,
Where the jaguars prowl and the lizards crawl
On a broken altar and sculptured wall,
Where Mayan rulers in arrogant pride
Dreamed and schemed and suffered and died.
The inlaid thrones and the sacred urns
Are filled with orchids and stag-horn ferns,
The witching moon of the tropic skies
Caresses the lips and the dead stone eyes
Of fallen idols of lust and blood
That lie in the mold and the reeking mud
Of fever-jungles. The dust and bones
Of men who quarried and laid the stones
Of fabulous cities are turned to earth.
The echoes of prayers and chants and mirth
Of vanished people and priests and kings
Are heard in the night-wind's whisperings.
The seas and straits and bays and coves,
The peaks and the valleys and swamps
    and groves
Hold ruins of dreams that were dreamed
    by men
In centuries lost and beyond our ken.
There are names that were spoken by
    long-dead lips
Of men who came in their gallant ships,
    Bahama . . . Barbados . . . Havana . . . Bermuda . . .
    Jamaica . . . Tortuga . . . Caracas . . . Barbuda . . .
    Martinique . . . Port au Prince . . . Santiago de Cuba . . .
    Windward Isles . . . Leeward Isles . . . Isle of
    Pines and Oruba . . .
Trinidad . . . Margarita . . . Tobago . . . Inagua . . .
Orinoco . . . Honduras . . . Yucatan . . . Nicaragua
Guatemala . . . Porto Rico . . . Costa Rica . . .
    Cartagena . . .
Venezuela . . . Baranquilla . . . Maracaibo . . . Magdalena.
Florida thrusts like a guiding thumb
To the southern islands of rumba and rum,
To the lands of mystery that lie below,
To the places I know I'm going to go.
                        —Don Blanding

# Contents

Plates viii

Figures ix

Foreword xi
    Richard Evans Schultes

Preface xii
    J.-P. Theurillat

Acknowledgments xiii

Map 1. The Caribbean Region xv

Map 2. Montserrat xvi

Introduction 1

Description of the Study Area 7

Ethnobotanical Uses and
    Specific Discussion 15

Appendix: A Plant Collection
    from Montserrat 135

References Cited 157

Index of Common Names 161

Index of Scientific Names 169

# Plates

*Following page 68*

1. *Plumeria rubra*, red frangipani
2. *Philodendron giganteum*, big chaney bush
3. Coconut (*Cocos nucifera*) inflorescence
4. Patricia Rabess with coconut leaf fibers
5. *Crescentia cujete*, calabash
6. *Bixa orellana*, annatto
7. Silkcotton dugout canoe
8. Informants with mammy apples
9. The author collecting beach morning glory specimens
10. *Momordica charantia*, maiden apple
11. *Sechium edule*, christophine
12. Informant with *Diospyros revoluta*, black apple
13. *Hura crepitans*, sandbox tree
14. *Caesalpinia bonduc*, grey nicker
15. *Piscidia piscipula*, dogwood
16. *Hibiscus rosa-sinensis*, hibiscus
17. *Hibiscus sabdariffa*, sorrel
18. *Artocarpus altilis*, breadfruit
19. Fred Payne with *Castilla elastica*, wild rubber
20. *Heliconia caribaea*, balisier
21. James Lee cutting sugarcane
22. *Coccoloba uvifera*, seagrape
23. *Mimusops coriacea*, Aphrodite's apple
24. *Datura suaveolens*, angel's trumpet

# Figures

1. A bush tea vendor   3
2. The public market   4
3. The precipitous rocky coastline   8
4. Prehistoric Amerindian artifacts   11
5. Amerindian zimi   12
6. Montserrat viewed from the sea   14
7. *Agave beauleriana*, coratoe   16
8. *Anacardium occidentale*, cashew   17
9. *Mangifera indica*, mango   19
10. Flowers of *Annona montana*, wild soursop   21
11. *Annona muricata*, soursop   22
12. *Annona squamosa*, sugar apple   23
13. Informants with *Anthurium grandifolium*, crackers   27
14. *Colocasia esculenta*, dasheen   28
15. Informants with *Xanthosoma sagittifolium*, calalu   30
16. *Acrocomia aculeata*, gru gru palm   31
17. James Cabey's method of climbing a coconut tree   33
18. James Cabey picking coconuts   33
19. Dehusking a coconut with a rock   34
20. An opened green coconut   35
21. *Phoenix dactylifera*, date palm   37
22. *Cordia obliqua*, clammy cherry   44
23. *Bursera simaruba*, gum tree, hosting *Phoradendron trinervium*   45
24. *Cephalocereus royenii*, dildo   46
25. *Melocactus intortus*, Turk's cap   47
26. *Opuntia cochenillifera*, cochineal cactus   48
27. *Canna edulis*, toloma, starch grains   49
28. *Manihot esculenta*, cassava   59
29. James Cabey grinding cassava   59
30. Informants with dried cassava meal   60
31. Eighteenth-century stone oven   60
32. Cassava bread drying   61
33. Informant cutting *Acacia farnesiana*   64
34. Charcoal pit   65
35. Thyme and banana companion planting   91

**Figures**

36. *Aloe vera*, sintibibi   94
37. Informant demonstrating ridge tillage   97
38. Boat and fish traps made of red cedar   99
39. Mahogany coffin suspended from a tree as a warning   101
40. *Cecropia peltata*, trumpeter tree   102
41. Hut thatched with *Heliconia caribaea* and *Musa sapientum*   104
42. *Syzygium malaccense*, Malay apple   107
43. *Passiflora edulis*, passionfruit   108
44. Abandoned sugar mill   111
45. Sugarcane crusher   112
46. Evaporating kettle   112
47. *Zanthoxylum monophyllum*, yellow Hercules   121
48. *Blighia sapida*, akee   121
49. *Datura innoxia*, wonga   124
50. *Solanum melongena*, white eggplant   127
51. Fruit of *Theobroma cacao*, cocoa   127
52. Seeds of *Theobroma cacao*   128

# Foreword

During the past half-century, ethnobotany has become established as a distinct interdisciplinary field of research. Recently, with the worldwide recognition of the urgent need to care for and preserve the environment, ethnobotanical conservation has come to be recognized as an important and valuable adjunct of environmental conservation. The knowledge of the bioactive properties of plants amassed over the millennia by peoples in primitive societies, whether they be Indians or simple country folk, is fast disappearing with increasing "westernization" in many, if not most, tropical regions. This precious knowledge is being lost sometimes more rapidly than the species of trees.

The publication of this book is a valuable contribution to the development of Caribbean ethnobotanical investigations. The Caribbean is a region that has not had sufficient ethnobotanical study, primarily because most of its original Indian peoples were systematically exterminated. Yet the present inhabitants, whatever their origin, have had to live on the useful plants of their environment and are no less interesting than the Indian populations that once inhabited the various islands before European annihilation of the natives.

Brussell's methods of field research, his citation of voucher specimens, his broad knowledge of the local flora, and his successful field work all bespeak his outstanding personality and unusually expert training in ethnobotanical investigation. It is to be hoped that a specialist so well suited and educated to carry out difficult ethnobotanical studies and field work will be able to dedicate his academic career to this field of research and in regions in the American tropics (for example, the Amazon) direly in need of qualified ethnobotanical research scientists.

Richard Evans Schultes, Ph.D., FMLS
Edward Charles Jeffrey Professor of Biology
Director, Botanical Museum of
   Harvard University (Emeritus)

# Preface

With the publication of *Potions, Poisons, and Panaceas*, David Eric Brussell makes a significant contribution to the ethnobotanical information available on the Caribbean island of Montserrat. His effective use of island informants allowed him to locate and study a greater number of plants than would have otherwise been possible, gave him access to a mine of information about local medical and domestic uses of the plants, and also introduced him to folkloric aspects of Caribbean culture, such as voodoo.

Brussell's careful observation of the medicinal uses of the plants led him to test on himself the efficacy of some treatments. His field research further supports scientific evidence that certain of the plants may have a future in the treatment of AIDS and cancer. Another important feature of the book is the fact that many of the plants studied come from diverse areas outside the Caribbean region. While most of the plants are from the tropics, some originate from the temperate regions of the Old and the New World. Thus, the book will permit the ethnobotanical comparison of the utilization of the same plant from one region of the world to another. This aspect is especially significant for the more than thirty economically important species that are widely cultivated.

And finally, it is not the least merit of the book that the many plant utilizations it documents enliven the sometimes austere ways of systematic botany teaching.

Dr. J.-P. Theurillat
Conservatoire Botanique
Genève, Switzerland

# Acknowledgments

Heartfelt gratitude is extended to Prem Rawat. I would especially like to thank Dr. Donald Ugent for his help and inspiration during the writing of this book. I am also grateful to Dr. Norman J. Doorenbos, Professor Richard Evans Schultes, Dr. Charles B. Arzeni, Dr. John W. Voigt, Dr. Robert D. Russell, Dr. Robert H. Mohlenbrock, Dr. Donald R. Tindall, Dr. Aristotel Pappelis, Dr. Walter J. Sundberg, and Dr. Isabella A. Abbott for their valuable assistance and suggestions.

I would like to thank Mr. S. P. McChesney, former president of the Montserrat National Trust, for his generosity in coordinating my research, housing, and transportation on the island. I also thank all other members of the Montserrat National Trust for their generous support and kindness.

Sincere appreciation is extended to Denis King, Franklyn Margetson, Annie Laurie, and Nymphus Meade for helping me locate informants. I am grateful to Mr. and Mrs. Alfred Payne, James Cabey, Daniel Allen, George Lee, Sara Lee, James Lee, Dan Daley, Sarah Rodney, Peggy Greer, James Corbet, Mary Gray, Mrs. Morson, Noel Pond, John McDaniel, Maria Ramdass, Eileen Beeson, Patricia Rabess, Nancy Dyette, and all the other kind people of this enchantingly beautiful island who generously served as informants.

Gratitude is extended to Dr. Lydia Pulsipher for providing me with contacts on Montserrat and for the valuable information she imparted prior to my departure for the Caribbean.

I would like to express my appreciation to Mr. and Mrs. S. P. McChesney, Denis King, Mr. and Mrs. Cedric Osborne of the Vue Pointe Hotel, Audrey Ciabattoni, and Mr. and Mrs. R. Marland for their hospitality in providing me with gracious living facilities and for helping me feel at home on Montserrat.

I express my gratitude to Dr. William Theobald, Director of the National Tropical Botanical Garden, Lawai, Kauai, Hawaii, for his assistance during my work on this project.

Much appreciation is extended to my grandmother, Esther Trout, my parents, Jean and William Brussell III, and to Kevin Brussell, my brother, for their

## Acknowledgments

love, help, and patience. I want to thank my mother for reading to me when I was a child and for helping me appreciate the joy of learning. I would like also to express my gratitude to my father for the vivid stories of his service in the United States Navy on an aircraft carrier in the South Pacific during World War II. His colorful accounts piqued my interest in the South Seas at an early age.

I would like to thank the following people for their generous assistance: Lydia Belle Byard, Chienfang Sari Ramsey, James D. Simmons, Susan H. Wilson, Sally Master, Sandy Tinsman, the Hollitz family, the Sailon family, the Foley family, Earl G. Bingly, Davey Coutts, Barbara Kolodney, Mr. and Mrs. Parsons, Mr. and Mrs. William White, Hugh Mitchell, Karen Schmitt, Dr. Kenneth A. Pavlicek, Bill Potter, Dr. Krishnamani, David George, Bob McChesney, Dr. Chris Lee, Gail Baumgardner, Leonard Abess, Jane Abess, Ruth McNab, Mr. and Mrs. Tommy Coulthard, Theodore G. Glass, Florence Glass, Anne Wood, Ruth Watters, Michael King, Ann Saunders, Dr. Benjamin A. Shepherd, Dr. John H. Yopp, John Richardson, John Vercillo, Dr. Francis J. Menapace, Dr. Jared H. Dorn, Dr. Charles B. Klasek, Beverly Walker, Terry Waite, Tom Vuke, Kenneth Miller, Niki Goulandris, Dr. Bryan Adams, Ron Liesner, Dr. J. J. Johnson, Dr. Charles Heiser, Dr. John Dwyer, David Breen, Dr. Frank Almeda, Dorothy L. Whitacre, Joanne Tabels, Kenneth Hayden, Lili Bartes, Dr. Alan Froehling, Barbara Froehling, Dr. Ralph Thompson, Linda Lane, Mr. and Mrs. Parr, Mr. and Mrs. Christopher Tyson, Justine Kalka, Brenda Cazewell, George Alexander, Dee Boatman, Stephen Hoffmann, Helen Vergette, Freddie Waddell, Su Clauson, Karen Stollmeyer, and Dr. Susan Ford. Some who may have been unintentionally left out of my acknowledgments are herewith given my thanks.

Thanks to Jimmy Buffett, the Beach Boys, and Arrow for all the great music.

This research was funded in part by grants from the Montserrat National Trust, the Southern Illinois University at Carbondale College of Science, the National Tropical Botanical Garden, and Sigmi Xi, the Scientific Research Society of North America. Their support is very much appreciated.

**Map 1.** The Caribbean region showing Montserrat (drawn by Karen Schmitt, courtesy of Dr. D. Tindall and Dr. S. Ford).

Map 2. Montserrat (redrawn by Tom Vuke and David Brussell from British Directorate of Overseas Surveys, 1967).

# Introduction

Montserrat is an island with a diverse flora and a rich heritage of plant folklore, the latter encompassing traditions passed down from Africans, Caribbean Amerindians, and Europeans. Much of this knowledge is still available in the older people of the island but would be lost if this information were not recorded while the elder generation is still living. This is particularly true because the younger generation does not seem to be interested in maintaining the older traditions.

With these ideas in mind, I began research on the ethnobotany of Montserrat. However, a review of the literature revealed that there were few botanical reports pertaining to Montserrat and that the flora of the island was poorly known. It then became clear that an ethnobotanical study that included scientific collecting of plant specimens, as well as consulting and observing informants and studying markets, might add considerable taxonomic information pertaining to the flora of Montserrat. The possibility of finding a plant that offered a new food, drug, insecticide, or industrial chemical source was also an incentive for pursuing this study.

## EARLY STUDIES

The history of botanical work on the island is summarized by Howard (1961). According to Howard, a Reverend Mr. Clark made a small collection of plants from Antigua and Montserrat in the years 1734 through 1736. This collection is now part of the Sloane Herbarium in the British Museum of Natural History in London. Following this, Dr. Patrick Browne made a collection of plants on Montserrat in 1760. His collection was afterward given to Dr. Edward Hill, professor of botany at the University of Dublin. A few of his plants are cited by Grisebach. During the latter half of the eighteenth century, Dr. John Ryan, a physician, made an important collection on the island that was sent to Professor Martin Vahl in Copenhagen. Fifty-two of his specimens were new to science.

Two other eighteenth-century botanists who visited Montserrat were Julius Philip Benjamin de Rohr and Joseph Dombey. Rohr was stationed on St. Croix as the director of agriculture during the 1780s, and in 1784 through 1785 he collected plants on Montserrat that were cited by Vahl (Howard 1961). Dombey, a French botanist who is known for his collecting in South America, was captured by pirates while leaving Guadeloupe and held on Montserrat by the British. He died in prison on Montserrat in 1796 (Coates 1969).

The first published work in the nineteenth century to include Montserrat was the *Flora of the British West Indian Islands* by Grisebach (1864). This book was the most comprehensive taxonomic treatise on the Caribbean area published in the nineteenth century. Grisebach described 3,143 species, but only forty-eight plants were cited from Montserrat (Howard 1961).

The Royal Botanic Gardens (Kew Gardens), London, has also had a part in the history of botanical exploration of Montserrat. In 1879 Kew Gardens received a collection of sixty Montserratian plants from a Reverend H. K. Holme. Later, in 1891, the assistant director at Kew Gardens, Mr. D. Morris, visited Montserrat and wrote an official report that described many common wild and cultivated plants growing in easily accessible areas of the island.

Two other nineteenth-century workers who visited Montserrat are Nichols and Barber. In 1891 Nichols, a medical worker from Dominica, made a trip to Montserrat to study the distribution of yaws (a disease of the tropics), during which time he collected some plant specimens. Barber collected plant specimens on Montserrat that were mostly cultivars and weeds in 1893. At that time, he was stationed on Antigua as the superintendent of the Agricultural and Botanical Department of the Leeward Islands.

J. A. Shafer (1907), a botanist with the New York Botanical Garden, was the first known American to do work on Montserrat. He made a large collection of dried plant specimens and published his findings in the *Journal of the New York Botanical Garden*.

## POTIONS, POISONS, AND PANACEAS

A taxonomic survey of the ethnobotanically important plants of Montserrat was conducted with the assistance of Montserratians who were highly knowledgeable in plant usage and folklore. Compiling the ethnobotany of Montserrat was a viable and productive endeavor. The island has a rich diversity of plant life, and raw plant materials were found to be very important in meeting the needs of everyday life of the inhabitants of Montserrat. While living on the island, I made many friends and took advantage of numerous opportunities to make observations on the ways plants were utilized. A great deal of information was obtained from interviews, especially on how plants are and have been used for medicine, food, fiber, poisons, dyes, building materials, and rituals. (The various plants relevant to the local voodoo folklore proved to be especially fascinating.) Informants were approached as persons of knowledge and authority. It was a pleasure to record the cumulative knowledge of many generations of these farmers, fishermen, bush doctors, herdsmen, gardeners, midwives, woodsmen, carpenters, and sailors. Notes were taken and some interviews were recorded with a small cassette recorder. I was pleased with the fact that so many Montserratians were happy to share their knowledge. Their contribution to science is a great gift.

In addition to collecting in the field, I also made weekly visits to the public market and the government market, both of which are located in Plymouth (figures 1 and 2). Observations were made as to what plants were sold as items of commerce, and some purchases were made in order to collect specimens. Also, I personally experimented with some plants in order to gain a deeper understanding of their ethnobotanical uses.

Literature describing plant uses in the West Indies was also consulted. However, very little has been written about plant usage on Montserrat. The works of Shafer (1907), Duberry (1973), and Pulsipher (1977) are three written sources that offered information pertaining to plant usage on the island. *Flora of the British West Indian Islands* by Grisebach (1864), *Flora of*

**Figure 1.** A bush tea vendor taking medicinal plants to the market.

**Figure 2.** The public market in Plymouth.

*Barbados* by Gooding et al. (1965), *Flora of the Lesser Antilles* by Howard (1974–1989), *The Bahama Flora* by Britton and Millspaugh (1962), and *Flora of the Bahama Archipelago* by Correll and Correll (1982) were utilized for taxonomic purposes.

As a result of this study, the first ethnobotanical collection from Montserrat was assembled. Study materials were placed in plant presses and dried with the aid of a propane gas oven. Voucher specimens were deposited in the herbarium of the Southern Illinois University Department of Botany and the herbarium of the Montserrat National Trust Museum. Some duplicate specimens were sent to the National Tropical Botanical Garden Herbarium in Hawaii, the Missouri Botanical Garden Herbarium, the U.S. National Herbarium at the Smithsonian Institution, the Gray Herbarium at Harvard University, the herbarium of the Royal Botanic Gardens at Kew, and the Stover Herbarium of the Eastern Illinois University Department of Botany.

A total of 282 ethnobotanically important plants are considered in this book. Of this total, 207 (73 percent) are medicinal, 123 (44 percent) are used for food, 49 (17 percent) are poisonous, 41 (15 percent) are a source of wood, 27 (10 percent) are associated with voodoo and folklore, 14 (5 percent) are sources of fiber, 9 (3 percent) are utilized for production of dyes, 8 (3 percent) are employed as aphrodisiacs, and 32 (11 percent) have various, miscellaneous uses which include hallucinogens, aromatics, insect repellants, ornaments, brooms, and teeth-cleaning agents.

My regional collections are designated by a capital letter preceding the specimen number. In this study the letter C designates specimens collected in the Caribbean.

The results of this study suggest that further pharmacological investigations of the various medicinal species known from Montserrat should be made. The various fish poisons, rodent poisons, and insecticides are also worthy of further study. Biochemical analysis of some of the more promising species is currently being performed. A new medicine, analgesic, antibiotic, noncarcinogenic insect repellent, or an effective, yet environmentally sound insecticide may come from one of these plants.

## CAUTIONARY NOTES

This research was performed for scientific purposes. Due to the poisonous properties of certain plants and the possibility of adverse reactions from some plants, I do not recommend employment of any ethnobotanical uses listed in this book and will not be held liable for the effects on anyone who employs them.

The plant descriptions in this book are included to give the reader an overview of the appearance of the plants, not to serve as a means of identification. Positive identification should be made only with the aid of an expert.

Furthermore, as experienced collectors know, field taxonomy is not so easy as it appears; there are often unexpected challenges to deal with. One of the health risks worthy of mention on Montserrat and certain other Caribbean islands is schistosomiasis (Lee 1980). Taxonomists and other field biologists must be aware of this situation. Field workers should avoid wading in questionable freshwater ponds or streams unless proper protective gear is worn. Care should also be taken to be sure of the safety of drinking and bathing water.

## CONSERVATION

It is imperative that we preserve the world's tropical rainforests including those of the West Indies. Maintaining biodiversity by preserving wild species is important for numerous reasons. A considerable amount of the world's free oxygen is produced by plants of the tropical rainforests where photosynthesis is not interrupted by winter. Many noteworthy medicines come from plants including the powerful anti-cancer drugs etoposide, teniposide, vincristine, and vinblastine. Other anti-cancer drugs and perhaps even a cure for AIDS may be present in a rainforest plant. Wild plants also offer new sources of food, industrial chemicals, environmentally safe pesticides, building materials, fibers, and alternatives to gasoline, diesel fuel, and oil. Tragically, some rainforest species are being exterminated before they can even be named.

The Montserrat National Trust has wisely proposed that tracts of land above 457 m be protected as natural areas. Some species of plants in the West Indies are becoming scarce. It is important that wild plants are not overcollected and that care is taken to assure continuing populations of the native plants of the Caribbean region. Ethnobotanical investigation can provide realistic support for conservation of endangered species and natural areas.

## RECENT ERUPTIONS ON MONTSERRAT

Recent eruptions of the volcano on Montserrat (see, for example, the July 1997 issue of *National Geographic*) have made the data collected for this book especially noteworthy. Many of the species were collected in the area devastated by the volcano. It is still too early to know how many of the plants can still be found on the island. Moreover, the deaths of elderly informants, each with his or her unique and irreplaceable knowledge of ethnobotanical data, have made the information contained in this book all the more precious.

small mangrove thicket near Hot Water Pond in Sand Ghaut. These dry woodlands are dominated by small *Acacia* trees, *Croton* shrubs, *Plumeria alba* L., *Cephalocereus royenii* (L.) Britton & Rose, *Melocactus intortus* (Miller) Urban, and species of *Opuntia*.

In the lower Olveston area, the Cudjoehead area, and on St. George's Hill, Caribald Hill, and Smokey Hill, there are seasonally dry forests that are intermediate between the interior montane forests and the xeric coastal woodlands. The dominants in these seasonally dry forests are: *Cedrela mexicana* M. J. Roem., *Hymenaea courbaril* L., *Bursera simaruba* (L.) Sarg., *Hura crepitans* L., *Pimenta racemosa* (Mill.) J. W. Moore, *Swietenia mahagoni* (L.) Jacq., *Inga laurina* (Sw.) Willd., and *Tabebuia pallida* (Lindl.) Miers.

As one ascends the higher slopes, above 305 m, tropical rain forest is encountered. At the lowest elevation within this zone a *Swietenia-Tabebuia-Haematoxylum* assemblage prevails.

On some of the intermediate moist slopes, extensive stands of *Cyathea arborea* (L.) J. W. Smith and groves of *Rhyticocos amara* (Jacq.) Beccari are interspersed with *Clusia rosea* Jacq., *Calophyllum antillanum* Britton, *Byrsonima crassifolia* (L.) Kunth, *Inga laurina*, and *Simarouba amara* Aubl. Due to intense cultivation practices in the past, much of the rainforest in this zone is secondary; however, patches of relatively undisturbed forest exist in some places. Beard (1949), for example, refers to "a patch of good *Dacryodes-Sloanea* rainforest, at 488–549 m, on the west slope of the South Soufriere Peak above O'Garra's." Similarly, I observed mature *Dacryodes-Sloanea* rainforest stands on the south slope of Castle Peak at about 579 m and in the Katy Hill area of the Centre Hills at about 549 m.

In addition to *Sloanea* sp. and *Dacryodes excelsa* Vahl, rainforest dominants near the upper level of the intermediate moist slopes of the island include *Beilschmiedia pendula* (Sw.) Hemsley, *Richeria grandis* Vahl, *Podocarpus coriaceus* L. C. Rich., *Ceiba pentandra* (L.) Gaertn., *Cecropia peltata* L., *Ficus trigonata* L., *Didymopanax attenuatum* (Sw.) March., *Ixora ferrea* (Jacq.) Benth., *Micropholis chrysophylloides* Pierre, and *Guatteria caribaea* Urban. *Phyllanthus mimosoides* Sw. makes up most of the ground cover in the upper level rainforest.

At about 610 m, one encounters a mossy palm brake dominated by *Euterpe globosa* Gaertn., *Cyathea arborea*, *Hibiscus tulipiflorus* Hook., *Sloanea dentata* L., *Richeria grandis*, and *Marila racemosa* Sw.

Above 700 m, an elfin woodland prevails. This elfin woodland is characterized by gnarled, low, tangled vegetation, mostly under five meters high, matted with lianas and festooned with epiphytic bromeliads and bryophytes (Beard 1949). The dominants in this uppermost vegetation zone include *Charianthus alpinus* (Sw.) R. Howard, *Freziera undulata* (Sw.) Willd., *Hedyosmum arborescens* Sw., and *Ilex sideroxyloides* (Sw.) Griseb.

The remaining vegetation type found on Montserrat is that found near the volcanic vents. Beard (1949) uses the term "fumarole" to describe plants such as these which border the soufrieres. The sulphurous gases emitted by these volcanic vents are fatal to most plants. Resistant plants that survive in these areas include *Ternstroemia peduncularis* DC., *Clusia alba* L., *Palicourea crocea* (Sw.) Roem. & Schult., *Cyperus ligularis* L., and *Pitcairnia angustifolia* Ait.

## HISTORY

Montserrat was first inhabited by Amerindians that migrated up the Caribbee island chain from the Venezuela area. However, the cultural identity of the first inhabitants is still in question. According to Fergus (1975), the Ciboney or "stone people" may very well have developed a culture on the island prior to the arrival of the Arawaks. Fergus (1975) describes a small figurine with white on red ornamentation that incorporated human and animal features found in the soil at Dagenham. It was identified as a "modelled adorno typical of the Saladoid-Barrancoid horizon" by Dr. Ripley Bullen of the Smithsonian Institution. This artifact can be dated between 500 B.C. and A.D. 500, and points to the possibility of pre-Arawak life on Montserrat.

Work performed by Watters (1980) has greatly augmented our knowledge of pre-Columbian archaeology on Montserrat.

It would appear that Montserrat was occupied by Saladoid groups, at least at Trant's, by about A.D. 200 or thereabout. It may have been initially occupied a century or two before. The conservative estimate is given because of the occurrence of polychrome painting. The presence of ZIC, WOR painting, D-shaped strap handles, tabular incised lugs, and modeled-incised heads suggest occupation could have taken place during early Saladoid. The Radio Antilles (MS-A1) ceramics indicate Saladoid occupation continued. The remaining sites confirm post-Saladoid occupation, although how long it persisted is uncertain. In general, the chronological framework for Montserrat, based mainly on ceramics, generally agrees with the regional chronology for the Leeward Islands and Lesser Antilles.

The Arawaks are thought to have arrived on Montserrat around A.D. 400 (Beard 1949; Fergus 1975). They practiced shifting cultivation and fishing and usually built their villages near the sea (Beard 1949). Excavations at Trants Estate have turned up figurines that may have been household gods, ceramic vessels, stone tools, and conch shell adzes that have been attributed to the Arawak culture (Fergus 1975) (figure 4). Zimis (believed by some scholars to be fetishes) have also been found on Montserrat (figure 5).

A few hundred years after the Arawak culture had been established, the Caribs began to move northward through the Caribbean Islands. By 1500, the fierce cannibalistic Caribs had preyed on and driven the peaceful Arawaks to the northern parts of the West Indies (Rouse 1946). The Caribs practiced both shifting and permanent cultivation and constructed houses made of pole framework and covered with leaf thatch. Buildings of this type are still being made in the West Indies. I observed structures of this type on the Carib reservation on Dominica during December 1978.

Carib villages and gardens were usually constructed near the coast. However, they also tended and harvested plants in the forest and maintained garden fortifications in the interior hills and mountains, where they could retreat when under attack (Beard 1949; Taylor 1938).

There is evidence of one of these interior gardens between the Soufriere Hills and St. George's Hill. The garden was described in a deed written in 1702 as "the Great Indian Garden" (Oliver 1917). This area is now a thicket known by some Montserratians as "The Garden" (Pulsipher 1977).

According to Smith (1630), Harris (1963), and Jesse (1966), the following plants were cultivated by the Caribs: *Indigofera suffruticosa* Mill., *Bixa orellana* L.,

**Figure 4.** Prehistoric Amerindian mortar and pestles for grinding cassava and other materials (with conch shell adze in left foreground) that were found on the island.

**Figure 5.** Prehistoric Amerindian artifact (known locally as a zimi) found on the island.

*Genipa americana* L., *Crescentia cujete* L., *Spondias mombin* L., *Ceiba pentandra* (L.) Gaertn., *Psidium guajava* L., *Ipomoea batatas* (L.) Lam., *Zea mays* L., *Mammea americana* L., *Arachis hypogaea* L., *Capsicum annuum* L., *Phaseolus vulgaris* L., *Phaseolus lunatus* L., *Ananas comosus* (L.) Merr., *Carica papaya* L., *Persea americana* Mill., and *Solanum tuberosum* L.

Alliouagana, the Carib name for Montserrat, meant "island of the prickly bush" (fide Fergus 1975). It has been suggested that this name refers to the aloe plant; however, it does not seem likely to me since aloe was brought to the West Indies by Europeans (Gooding, et al. 1965). Alliouagana probably refers to one or more of the native species of *Acacia*.

## Description of the Study Area

Although Montserrat was sighted and named after a mountain in Spain by Christopher Columbus in 1493, it was not settled by whites until settlement by the British in the seventeenth century (Fergus 1975) (figure 6). A vivid impression of Montserrat and the surrounding area was described by Sir Henry Colt in 1631:

> Weddensday. 20. July. We arriued att Montserrate, y$^e$ land high rownd montaynous & full of woods, w$^{th}$ noe Inhabitants; yett weer y$^e$ footstepps seen of some naked men.
>
> In these. 4. dayes space I neuer felt soe moyst an ayre. All things rust, y$^e$ verye keyes in our pocketts rust, & a nights y$^e$ clothes of our backs in touch is moyst, & stiff. We are moor drowsye and sleepy then accustomed, & full of dreams. We would willingly haue beaten it further vpp vnto Antigo, beinge butt. 7. leagues of, a place of moor securitye; but y$^t$ there is no water ther.
>
> We approchinge thus y$^e$ harbar of Montserrate, we send vpp to y$^e$ topp of our mayne mast to discouer harbour land and sea befoor we would lett fall our Anchors. All day we keep our watch vppon y$^e$ mayne mast; butt all beinge cleer we come to our Anchor on y$^e$ weast side of Montserrate, in sight of Rodunda our next Iland, beinge noe other then a single rock. Itt is high rownd & nothinge els but one stone. (Harlow 1924)

In 1632 the British sent Irish Catholic dissidents from St. Kitts to colonize Montserrat (Montserrat National Trust 1976). After initial settlement, Montserrat became an asylum for political and religious refugees from other colonies. Many political prisoners from Ireland were transported to Montserrat (Fergus 1975; Pulsipher 1977). Scotch settlers also came to Montserrat, however, the Irish far outnumbered them.

The colonists of Montserrat grew "large quantities of tobacco and sugar" according to Colonial Office records of 1654. Cultivation of sugarcane led to the procurement of black slaves whose origin was Africa. By 1672 there were 523 slaves on the island. Slaves continued to be the main labor force until their emancipation in 1838 (Fergus 1975). The freed blacks began their own subsistence farms and worked in various capacities as laborers for plantation owners (Pulsipher 1977). After 1850 limes were an important export crop for the island. Black Montserratians played a major role in the cultivation of these three important crops (Fergus 1975).

Although there are no Caribs presently living on the island, it is believed that some intermarriage between Caribs and blacks occurred in the past.

The present population of Montserrat is primarily of black African descent. There is also a small minority of caucasian expatriots from the United States, Canada, Great Britain, and other European countries, who reside on the island. Some individuals of East Indian and Middle Eastern descent also now live on the island.

In 1973 Montserrat had a population of about 12,500, one fifth of which lived in Plymouth, the capital (Irish 1973). The 1977 mid-year population estimate for Montserrat was 12,160 (Watters 1980). The new Rand McNally College World Atlas states that the population of Montserrat is 12,000 with 3,000 people living in Plymouth (*New Rand McNally College World Atlas*, 1983).

**Figure 6.** The west coast of Montserrat viewed from the sea. Columbus may have seen a similar view when he sighted and named Montserrat in 1493.

# Ethnobotanical Uses and Specific Discussion

## ACANTHACEAE

*Justicia pectoralis* Jacq.; Bitter Balsam, Garden Balsam

This erect herb, with opposite narrow-lanceolate leaves and small white and purple flowers, has short pointed capsules bearing small lens-shaped seeds. It is native to tropical America and the West Indies. *J. pectoralis* was collected at the market in Plymouth. Brussell C-347.

The dried leaves and stems are used to make a "bush tea" that is imbibed as a treatment for colds. This plant is in enough demand for bush tea to cause it to be cultivated for sale in the markets. In Amazonia, the plant is believed to be a treatment for flu and pneumonia, according to Professor Richard Evans Schultes of Harvard University.

*Ruellia tuberosa* L.; Double Bit, Monkey Gun, Many Roots

This suffrutescent herb has four-angled stems, tuberous roots, opposite obovate leaves, large purple flowers, and a cylindrical elongated capsule bearing several small seeds. This plant is found in the West Indies and northern South America. It was collected at Woodlands, St. George's Hill, and Cudjoehead. Brussell C-140, C-321, C-336.

The leaves are boiled, and the resulting decoction is drunk to treat gonorrhea, stomach pains, and colds.

## AGAVACEAE

*Agave beauleriana* Jacobi (= *Agave franzosini* J. G. Baker); Coratoe, Maypole; figure 7

This large fleshy herb has obovate-lanceolate, concave, basal leaves with marginal prickles and a basally grooved apical spine. The large yellow flowers are borne in panicles on a tall erect peduncle and produce elongate-claviform capsules bearing shiny black seeds. The plant is native to tropical America, but has been planted in other warm areas of the world. It was found at Shoe Rock. *Sight record.*

The juice squeezed from the sliced leaves is applied to the scalp to get rid of dandruff.

### *Agave sisalana* Perrine; Sisal

This large fleshy herb has linear-lanceolate, flat, basal leaves, each tipped with a brown apical spine. The large yellow-green flowers are borne in panicles on a tall erect peduncle and give rise to oblong capsules. This native of tropical America has been widely planted in the warmer areas of the world. It was found at Cudjoehead. *Sight record.*

The fibers extracted from the leaves of this plant are used locally to make twine and rope. This use was much more common in the past than at present.

Figure 7. *Agave beauleriana*, coratoe.

## AMARANTHACEAE

### *Achyranthes indica* (L.) Mill.; Man Better Man, Hug-Me-Close, Soldier Rod, Devil's Horse Whip

This erect herb has opposite obovate-rotund leaves and small inconspicuous flowers that produce fruits encased in spine-tipped bracts, which easily attach themselves to passing people or animals. It is a common dry-pasture weed in warm areas of the world. It was collected in a pasture at Woodlands. *Brussell C-141*.

Tea made from the leaves is a popular cold medicine.

*Amaranthus dubius* **Mart. ex Thell.; Spinach**

This erect herb has alternate rhombic-ovate leaves and small green flowers that produce small utricles. *A. dubius* is found in tropical America and tropical Africa. It was collected in a dooryard garden in Salem. *Brussell C-15, C-234*.

The leaves are eaten fresh in salads and cooked as a potherb.

## ANACARDIACEAE

*Anacardium occidentale* **L.; Cashew; figure 8**

This tree has alternate elliptic leaves and fragrant, showy, pink flowers that give rise to a large curved nut borne on a fleshy red receptacle. *A. occidentale* is native to tropical America, but is now widely grown in tropical areas of the world. It was found near the White River. *Brussell C-48*.

A decoction of the bark is used to treat diarrhea, diabetes, and articular swelling due to syphilis. The toxic, highly caustic oil from the pericarp of the nut is employed topically to treat skin ulcers, infections, and herpes lesions and as a topical agent for removing warts and corns.

**Figure 8.** *Anacardium occidentale*, cashew.

The swollen, fleshy, red or yellow receptacles of the nuts are edible and are referred to as "cashew cherries" on Montserrat. They are said to be a good treatment for an upset stomach when eaten fresh. I found these juicy sweet-tart "cherries" to be a refreshing snack while collecting in the dry areas of the island.

The cashew seeds are edible raw or roasted. When cracking open the raw nuts, one must wear gloves or otherwise exercise caution in order to prevent the caustic cardol oil from coming in contact with bare skin, or blisters will result. Roasting the nuts removes the poisonous oil; however, the caustic fumes may blister the skin and cause inflammation of the eyes and mucous membranes. Some individuals who come in contact with the leaves or other parts of this relative of poison ivy (*Toxicodendron radicans* L.) develop localized tissue reactions that may become systemic.

While exploring in the area near Great Alp Falls, I encountered a cashew nut roasting station consisting of a fire pit and three large, flat rocks.

The nuts and receptacles seem to be used only for local consumption. The nuts are not exported from the island, and no nuts or cherries were observed in the markets. Given the favorable climate and land available, along with the light weight and high price of these nuts, it seems growing cashews would be a profitable enterprise that could stimulate the island's economy and create new jobs.

*Comocladia dodonaea* (L.) Urban; Christmas Bush

This small evergreen tree has alternate spine-bearing leaves and tiny red flowers that produce elliptical drupes. It is native to the West Indies and was observed in the White River Valley. *Sight record.*

The spiny evergreen foliage and the bright red fruits give this plant an appearance similar to holly (*Ilex opaca* Ait.) and account for its use as a Christmas tree. The sap has been used as a black dye, but it must be handled with caution due to its caustic properties. Some people develop a skin rash after coming in contact with this plant.

*Mangifera indica* L.; Mango; figure 9

This large tree has alternate oblong-lanceolate leaves and numerous small yellow to purplish flowers that give rise to large ovoid drupes. This native of tropical Asia has been widely cultivated and sparingly naturalized in tropical areas of the world. It was collected at Salem and in a dooryard garden at Woodlands. *Brussell C-117, C-293.*

A tea made from the leaves is imbibed to treat colds, sinus congestion, asthma, coughs, and arthritis. The leaves are chewed to expel intestinal worms, but merely handling the leaves can cause skin irritation in some people. The resin from this tree is said to have antisyphilitic properties. A decoction of the bark or the dried pulverized seeds is taken internally for chronic diarrhea.

Mango is probably the most popular fruit on Montserrat. The sweet juicy fruits are usually eaten raw, but they are also cooked in chutney (a condiment) and pies, added to ice cream, and made into drinks.

Mango trees have escaped from cultivation in many areas of the island, and these so-called wild mangoes often yield small fruits with stringy fibers that get caught in one's teeth. These mangoes are sucked via a hole at one end but this may cause a rash around the mouth. They are best pressed for juice. There are accounts of children getting intestinal blockage as a result of a build-up of these fibers.

The leaves and bark of the mango tree are used to make a yellow dye for coloring cloth. Though the sap of the tree may be an irritant, the dried wood is used for chopping blocks, furniture, general construction, and boxes.

Some people are allergic to this poison ivy relative and develop a skin rash or other symptoms after coming in contact with it.

### *Spondias cytherea* Sonner; Golden Apple, Governor's Plum

This tree has alternate pinnate leaves with ovate leaflets and small white flowers that produce ovoid yellow drupes. It is native to the Society Islands in the South Pacific and has been cultivated in many tropical

Figure 9. *Mangifera indica*, mango.

regions of the world. It was found at Olveston. *Brussell C-252.*

The fruits are eaten raw or cooked for preserves.

### *Spondias mombin* L.; Hog Plum, Gully Plum, Yellow Mombin

This tree has alternate pinnate leaves with ovate leaflets and small yellow flowers that give rise to ovoid yellow drupes. *S. mombin* is native to tropical America. It was collected at Olveston. *Brussell C-253.*

The aromatic yellow fruits are eaten raw or are boiled to make a refreshing beverage. Hog plum jam is popular on the island and has a very fine taste and bouquet. The astringent bark is used for tanning leather. The light wood is used to make bottle stoppers and crates. Sections of the trunk are planted as living fence posts.

### *Spondias purpurea* L.; Jamaica Plum, West Indian Plum

This medium-sized tree has alternate pinnate leaves made up of ovate leaflets. The showy red flowers produce fleshy reddish to purple drupes. *S. purpurea* is native to the West Indies. The fruits were observed for sale in the Plymouth public market. *Brussell C-66.*

The fruits are eaten raw or cooked in pies and preserves.

## ANNONACEAE

### *Annona cherimola* Mill.; Dog Apple, Cherimoya

This small tree has alternate oblong leaves and yellow velvety flowers that give rise to large fleshy fruits. It is native to continental tropical America and has become naturalized in the West Indies. It was observed growing semi-wild on a hill above Salem. *Sight record.*

Many Montserratians enjoy eating the somewhat tart fruit and also make a tea from the flowers to treat diarrhea. The poisonous seeds are pulverized, and the resulting powder is used as an insecticide.

### *Annona glabra* L.; Jumbie Apple, Pond Apple, Monkey Apple

This small tree has alternate oblong leaves and axillary red and white flowers that produce aromatic, yellow, globose, aggregate fruits. This plant is found in the West Indies, the Atlantic coast of Mexico, Central America, and South America, as well as the coast of western Africa. *A. glabra* was found growing near the Emerald Isle Hotel. *Sight record.*

Tea made from the fruits, leaves, and twigs is imbibed to treat asthma. The fruits are sometimes eaten raw but have very little taste.

## Ethnobotanical Uses and Specific Discussion

***Annona montana*** Macf.; Wild Soursop; figure 10

This small tree has alternate oblong leaves and solitary, yellow, tomentose flowers that give rise to fleshy aggregate fruits containing shiny brown seeds. The wild soursop is native to the West Indies and is found throughout the Caribbean area. It was observed growing in a hillside thicket above Salem. *Brussell C-175*.

Tea made from the leaves, green fruits, and twigs is imbibed to treat asthma.

The bland yellowish pulp of the fruit is edible, but has very little flavor, and is not a popular food item. The light, soft wood is used by fishermen on the island to make floats for fishing nets.

The seeds are poisonous, as are those of all *Annona* fruits.

**Figure 10.** Flowers of *Annona montana*, wild soursop.

***Annona muricata*** L.; Soursop; figure 11

This compact tree has alternate oblong leaves and fleshy yellowish-green flowers that produce very large, green, irregularly conical, fleshy fruits. The soursop is native to tropical America. It was found growing in a hillside garden at Woodlands and at Plymouth. *Brussell C-109*.

Montserratians have found that when soursop branches are hung around the windows and doors of dwellings they are effective for repelling mosquitoes (Duberry 1973). A somniferous tea is made from the

Figure 11. Soursop (*Annona muricata*) fruit showing pulp and seeds.

leaves. The tea is taken before bedtime and is said to put a child to sleep in two minutes. In addition to its tranquilizing effect, the tea is used to treat menstrual discomfort and flatulence. An infusion made from the roots of this tree is imbibed to treat fish poisoning that results from eating fish that have a build-up of ciguatera toxin.

The ripe fruits are eaten raw or used in making drinks that, according to Montserratians, are effective for "cooling the blood" (lowering blood pressure) and for reducing fevers. Soursop drink is prepared by mixing water, milk, soursop pulp with the poisonous seeds removed, honey or sugar, essence of angostora, and butter. Preserves and ice cream are made from the fruit, and I found soursop ice cream to be particularly tasty.

The fruits and leaves are sold in the market, and both the beverage and the ice cream made from the fruit are very popular on the island.

The inner bark fiber taken from the trunk and large limbs is used to make rope and twine.

Powder made by grinding the poisonous seeds is used for insecticidal purposes on Montserrat.

### *Annona reticulata* L.; Custard Apple

This small tree has alternate oblong leaves and green and purple flowers that produce large, globose, brown or sometimes yellowish and red-blushed, fleshy fruits. This tree is native to tropical America and is widely cultivated throughout the tropics. It was observed growing at Woodlands. *Sight record.*

The custard apple tree is highly valued on Montserrat for its edible fruits, which are commonly eaten raw as well as in preserves and puddings. Custard apple fruits are often sold in the public market. The people of the island make an astringent tonic tea from the bark of the twigs and also use the bark as a source of strong fiber. The pulverized poisonous seeds are made into an insecticide that kills lice.

*Annona squamosa* L.; Sugar Apple, Sweetsop, Applebush; figure 12

This small tree has alternate elliptic leaves and fleshy green and purple flowers that give rise to fleshy, globular, tuberculate, green or bluish fruits with white, very sweet, seedy pulp. The sugar apple is native to the West Indies. It was observed growing in an upper level garden above Salem and collected at Olveston, Trants Estate, and in the White River Ghaut. *Brussell C-42, C-197, C-349.*

A decoction made from the leaves is taken internally to treat a cold, to get rid of "slime on the stomach," and "to cool the blood" (to reduce high blood pressure). Some people on the island employ the unripe fruits as an insect repellent, and the seeds are used for insecticidal purposes.

The sweet fruits are eaten raw and used to prepare

**Figure 12.** Fruits of *Annona squamosa*, sugar apple.

preserves and drinks. Sugar apple fruits can be found in the Montserrat market, and the drink made from the fruit is sometimes sold in village groceries and bars.

## APIACEAE

*Anethum graveolens* L.; Dill, Dillweed

This erect herb has alternate leaves that are pinnately divided into thin ribbons and small yellow flowers borne in umbels that give rise to dry, flattened, oblong fruits. Dill is native to Europe. It was collected in a dooryard garden on St. George's Hill. *Brussell C-309.*

The seeds are steeped, and the resulting tea is used to treat colds and fevers. Dill seeds are used as a flavoring agent in cooking also.

*Eryngium foetidum* L.; Ramgoat Bush, Fitweed

This biennial herb has a rosette of lanceolate basal leaves and alternate stem-borne leaves. The flowers are borne in cylindrical heads, and the dry compressed fruits are covered with scales. It is found in the West Indies and continental Tropical America. It was found growing in the White River Valley. *Brussell C-22.*

Tea made from the entire plant is imbibed for colds, pain in the bowels, and as a general conditioner. This tea is also sipped as an aphrodisiac and to make childbirth easier. Ramgoat bush was seen for sale in the public market in Plymouth as a bush tea aphrodisiac.

*Pimpinella anisum* L.; Anise

This erect herb has alternate simply divided pinnate to ternately divided leaves and umbels of yellow flowers that give rise to small, ribbed, ovate fruits. Anise is native to the Mediterranean area and has escaped from cultivation in many parts of the world. It was found growing in a dooryard garden at Salem. *Sight record.*

Tea made by boiling the seeds is administered as a febrifuge on Montserrat. Tea made from both the roots and seeds is sipped to counter hoarseness and indigestion. A stronger decoction of the same is drunk in one cupful doses three times a day as a vermifuge and emetic. One informant claims this plant is an ingredient in a love potion.

The chopped fresh leaves are added to salads and soups. The seeds are used as a flavoring agent in baked goods, sauces, and stews. Bundles of fresh anise stems and dry seed-bearing umbels were found in the public market at Plymouth.

## APOCYNACEAE

*Carissa macrocarpa* (Ecklon) A. DC.; Natal Plum

This dense shrub has opposite dark green leathery leaves, branches armed with two-pointed spines, and solitary fragrant white flowers that produce ovoid dark red drupes. The natal plum is native to South Africa and has been widely planted in the warm parts of the world.

This plant was collected in a dooryard garden at Salem. *Brussell C-233*.

On Monserrat the fruits are eaten fresh, cut up in salads, and used in sauces and preserves. The flavor is similar to that of black raspberries.

*Catharanthus roseus* (L.) G. Don; Dr. Dyette, Madagascar Periwinkle, Twelve O'clock, Everyday Flower

This erect herb has opposite leaves and showy white or pink flowers that produce narrow ribbed follicles. This plant is native to Madagascar but has been widely naturalized in tropical areas of the world. It was collected in a dooryard garden at Woodlands and at Plymouth. *Brussell C-34, C-322*.

A tea made from the leaves and roots is imbibed to treat strains, cancer, diabetes, and high blood pressure. Antitumor and antileukemic principles from the leaves, including vinblastine and vincristine, are now in wide use.

*Ervatamia cumingiana* (A. DC.) Markgr.; Milky

This glabrous green-stemmed shrub has opposite elliptic leaves and white flowers with salver-shaped corollas and produces finely ribbed winged follicles. It is native to the Old World tropics. Milky was found in the White River Valley. *Brussell C-165*.

The leaves and stems are broken and the white exudate is applied to the skin to cure ringworm.

*Nerium oleander* L.; Oleander

This large bushy shrub has whorled lanceolate leaves and fragrant red, white, or pink flowers that produce cylindrical finely ribbed follicles. Oleander is native to the Mediterranean region. It was collected in a dooryard garden at Woodlands. *Brussell C-229*.

The sap of this poisonous plant is used topically to treat ringworm and is added to grain and used to poison rats. A decoction made from the bark and leaves is drunk as a treatment for epilepsy and malaria. The powdered leaves are used for insecticidal purposes. Contact with any part of this plant can cause localized tissue reactions that may become systemic in some people.

*Plumeria alba* L.; Wild Frangipani, Milktree, White Frangipani

This small tree has alternate oblong leaves and fragrant white flowers that produce cylindrical follicles. *P. alba* is native to the West Indies. It was collected at Woodlands. *Brussell C-153*.

The toxic, white, thick latex from this plant is used to remove warts and as a balm to heal cuts and skin eruptions and to treat venereal disease. This white exudate is also used as a styptic. Montserratians use a tea made from the flowers as an expectorant. The leaves and bark are employed to make a decoction that is taken judiciously as a drastic purgative and as a treatment for syphilis, dropsy, and skin diseases.

The attractive pleasantly redolent blossoms are used to make leis, diadems, and other floral decorations.

**Plumeria rubra** L.; Red Frangipani; plate 1

This small tree has alternate obovate leaves and fragrant red or purple flowers that produce ovoid follicles. Red frangipani is native to the West Indies. It was collected at Woodlands and at Bugby Hole. *Brussell C-152*.

A tea made from the toxic roots and bark is used as a drastic purgative and also for treating syphilis and dropsy. A decoction of the bark is applied topically to treat skin diseases and relieve itching.

The fragrant showy flowers are strung into garlands and worn on special occasions.

**Rauvolfia nitida** Jacq.; Bitter Ash, Bitter Bush, Milk Bush

This small tree has whorled narrowly elliptic leaves and cymes of small white flowers that give rise to small reddish-black berries. *R. nitida* is native to the West Indies. It was found at Woodlands. *Sight record*.

The ground-up wood and bark are soaked in water for one or two minutes, and the resulting infusion is imbibed to treat high blood pressure, diabetes, and chest pain. The ground-up wood and bark are also used to poison or stun fish.

**Thevetia peruviana** (Pers.) K. Schum.; Luckynut, Luckyseed

This small tree has alternate linear leaves and showy yellow flowers that produce falcate drupes. *T. peruviana* is native to South America. It was collected at Salem. *Brussell C-158*.

A decoction made from the bark is sipped as a febrifuge and as an abortifacient.

The seeds and bark of this poisonous plant are pulverized and employed to kill insects and to stupefy fish. Children have been poisoned as a result of swallowing the toxic seeds. The seeds are carried as lucky charms and used to make necklaces. A Montserratian voodoo custom consists of putting the seeds with money to bring good luck in financial matters.

## ARACEAE

**Anthurium grandifolium** (Jacq.) Kunth; Crackers, Hard Leaf; figure 13

This robust, climbing, epiphytic or terrestrial plant has large, leathery, long-petioled, deltoid-cordate leaves with an open sinus or overlapping lobes at the base and an acuminate or cuspidate narrowed apex. The long-peduncled spadix is as long or longer than the leaves and two to three times longer than the linear-lanceolate spathe. *A. grandifolium* is native to the American tropics. It was found growing at Paradise. *Sight record*.

**Figure 13.** Informants gathering crackers (*Anthurium grandifolium*) to feed their pigs.

The leathery large leaves are utilized to wrap food and other items, as temporary umbrellas, and as food for goats and pigs.

### *Colocasia esculenta* (L.) Schott; Eddo, Dasheen, Eddee, Taro; figure 14

This robust herb has alternate sagittate-peltate leaves, an erect spadix, and a clasping spathe. It is native to India. Eddo was collected at Woodlands. *Brussell C-118, C-119, C-122.*

The corms of *C. esculenta* are peeled while wearing gloves (due to the irritating properties of the sap) and then boiled or baked and eaten. All parts of this plant contain a skin irritant.

### *Dieffenbachia seguine* (Jacq.) Schott; Dumb Cane

This erect herb has alternate ovate leaves, a white spadix, and a pale green spathe that bears red berries. Dumb cane is widely distributed in tropical America. It was collected in a ghaut above Salem. *Brussell C-297.*

Pieces of the roots, stems, and leaves are boiled with bait and used as rat poison. The sap is used to stupefy fish. The crushed leaves and stems are rubbed on sugar cane and various fruits to stop thieving. This plant has been used in voodoo practices on Montserrat.

Sections of the stems or petioles are dangerously used to induce abortion. Due to the poisonous properties of this plant and the physical damage that may

**Figure 14.** *Colocasia esculenta,* dasheen.

occur to the patient, I do not recommend induction of abortions with dumb cane stems. This practice could result in the death of the patient.

One informant stated that ingestion of the irritating sap causes temporary sterility in males. This too is hazardous.

*Monstera adansonii* Schott (= *Monstera pertusa* [L.] DeVriese); Sarsaparilla

This scandent vine has large, ovate, membranaceous leaves that are perforated along the midrib by oval holes. The spathe is yellow and about twice as long as the spadix. This species is native to the West Indies. It was collected in a fence row at Paradise. *Brussell C-195.*

A decoction made from the entire plant is applied topically to treat strains. The leaves are heated and used for making poultices.

This plant is not to be confused with *Smilax cumanensis* Humb. & Bonpl.; (wild sarsaparilla).

*Monstera deliciosa* Liebm.; Monster

This large vine has alternate large perforated and incised leaves, a deciduous cymbiform spathe, and an erect tough-stemmed green spadix which matures into a compound fruit with edible flesh. *M. deliciosa* is native to tropical America. It was found growing at Olveston. *Sight record.*

A decoction of the leaves and rhizomes is taken in a dosage of four cups daily to treat arthritic pain.

When the cone-shaped spadix is mature and the segments of "rind" are separating and falling off, the small sections of flesh are edible, sweet, and pineapple-like. These fruits should only be consumed when they are fully mature, for the crystals of calcium oxalate in unripe fruits cause a burning irritation in the mouth and throat.

*Philodendron giganteum* Schott; Big Chaney Bush; plate 2

This robust epiphytic or terrestrial herb has huge cordate-ovate leaves with long petioles, a convoluted spathe, a stout sessile spadix, and yellowish-orange fruits. The species is native to the West Indies. It was found at the Tar River Estate. *Sight record.*

The gigantic leaves are used instead of wrapping paper, as umbrellas, and as food for pigs and goats.

*Xanthosoma brasiliense* (Desf.) Engler; Chaney Bush, Crackers

This robust herb has alternate, sagittate to hastate, very large leaves, and a clasping yellow spathe shielding a spadix that bears many fleshy berries. This plant is native to tropical America. It was found near Great Alp Falls. *Brussell C-237.*

"Dotma" is the name given to sweet potatoes that have been wrapped in the leaves of this plant and cooked. The preboiled young leaves of this plant are used in the well-known calalu soup that is popular on Montserrat and other West Indian islands (cf. Hodge and Taylor 1957). The mature leaves are used as a protective wrapping for food and other items as well as for temporary umbrellas. The large leaves are also collected and fed to goats and pigs.

***Xanthosoma sagittifolium*** **(L.) Schott; Calalu; figure 15**

This aroid is a robust herb with large sagittate-ovate leaves borne on erect thick stems that arise from a starchy, swollen, underground structure. The oblong-ovoid greenish-white spathe is convoluted at its base and longer than the spadix. This plant of uncertain origin is cultivated in tropical areas. It was found in a dooryard garden near Salem. *Sight record.*

This plant is an important food source on the island.

**Figure 15.** Informants with *Xanthosoma sagittifolium*, calalu.

## ARECACEAE

***Acrocomia aculeata*** **(Jacq.) Lodd. ex Mart.; Gru Gru Palm; figure 16**

This stout prickly palm has a crown of very large, arcuate, pinnate leaves and glabrous drupes. Heretofore this tree was only reported from Dominica south through the Windward Islands to Grenada. However, Montserrat can now be added to its range. It was found at Woodlands. *Sight record.*

# Ethnobotanical Uses and Specific Discussion

**Figure 16.** Fruit of *Acrocomia aculeata*, gru gru palm.

The white oily flesh of the fruits is very similar in texture and taste to that of coconut and is eaten in the same way. Oil can be extracted from these fruits, and the hard shells are sometimes used to make rings.

### *Acrocomia media* O. F. Cook; Thorn Palm, Prickly Palm

This prickly tree has a crown of large arcuate pinnate leaves, a long spathe, and a spadix bearing yellow pistillate flowers near the base and staminate flowers at the apex. The fruits are yellow globose drupes. This species has previously been reported only from Puerto Rico, St. Thomas, and St. Croix. It was found at Woodlands. *Sight record.*

The white oily flesh of the fruits is eaten in the same way as coconut meat.

### *Areca catechu* L.; Areca Nut, Betel Palm

This graceful tree has large alternate pinnate leaves, small white flowers, and elliptic red drupes. This palm is native to tropical Asia but has been planted in various tropical areas of the world. It was found in the Salem area. *Sight record.*

Tea made by boiling pulverized pieces of the leaves and fruits is imbibed to get rid of intestinal worms and as a treatment for stomachache, diarrhea, and urinary disorders.

### *Cocos nucifera* L.; Coconut; figures 17, 18, 19, and 20; plates 3 and 4

This common tree has very large pinnate leaves, an erect boat-shaped spathe, and a branched spadix with male flowers at the apex and female flowers near the base. The fruits are large ovoid drupes that turn yellow or brown when ripe. The original home of this palm (long undetermined) is now stated to be Indonesia and the western Pacific islands. However, South America is the home of most of its related species. Coconut trees were commonly observed near most of the settlements on Montserrat. *Sight record.*

Because of its many uses, the coconut palm is considered an important ethnobotanical resource. Coconut meat is eaten raw and cooked. The soft meat of green coconuts is called "jelly" and has a consistency similar to commercial gelatin desserts. This sweet coconut jelly is scooped out of the freshly cut green coconuts with a spoon cut from a piece of the husk.

Coconut liquid endosperm (known as "coconut milk" when obtained from ripe coconuts and "coconut water" when taken from green coconuts) is a popular beverage. It is also fed to babies as an important source of nourishment. On a hot day the liquid endosperm extracted from a fresh-picked coconut is noticeably cooler than the surrounding air. Liquid endosperm from coconuts is very rich in nutrients and is used as a growth medium in some tissue culture methods em-

Figure 17. Informant James Cabey's method of quickly scaling a coconut tree (*Cocos nucifera*) is the most efficient on the island.

Figure 18. James Cabey picking coconuts.

**Figure 19.** James Cabey employing the traditional local custom of dehusking a coconut with a rock.

**Figure 20.** An opened green coconut, which is a source of coconut water (liquid endosperm) and jelly (partially developed coconut meat). The liquid endosperm was given intravenously to soldiers suffering from blood loss during World War II, according to one informant. A spoon cut from the husk (used to eat jelly) is to the left of the knife blade. Carolus Linnaeus wrote in the eighteenth century: "Man *dwells* naturally within the tropics and lives on the fruit of the palm tree. He *exists* in other parts of the world and there makes shift to feed on corn and flesh."

ployed in modern biotechnology. One informant stated that, in emergencies, liquid endosperm from green coconuts was administered intravenously to soldiers suffering excessive blood loss during World War II. Since the exact composition and properties of this rich mixture of nutrients and growth regulators are still unknown, research is being performed to learn more about the constituents of coconut liquid endosperm and their effects on living things.

Sap collected by tapping a branch of the inflorescence is gently boiled to produce an exquisite coconut syrup that is a local delicacy.

Coconut husk fiber is used to stuff pillows and cushions as well as to scrub pots and pans. The shells are used to make bowls, cups, spoons, ornaments, and flower pots. The leaves are woven into fans and mats, and the tough fibrous leaflet midribs are used to make brooms.

### *Euterpe globosa* Gaertn.; Cabbage Palm, Mountain Palm, Mountain Cabbage, Sierra Palm

This tree has large pinnate leaves, a glabrous spathe, and a branched spadix bearing white flowers that give rise to reddish-black drupes. This palm is native to the West Indies. It was collected at Woodlands and on Chance Peak. *Brussell C-133, C-323.*

The terminal bud is cut out and eaten fresh as "palm cabbage" or cooked in various dishes. I found this palm cabbage to be a refreshing source of nourishment and water while on collecting trips in the mountain rainforests.

The leaves are used to thatch roofs and weave mats.

### *Phoenix dactylifera* L.; Date Palm; figure 21

This tree has a crown of very large pinnate leaves, an erect spathe, and a branched spadix bearing cream-colored flowers that give rise to brown drupes. This Old World native has been widely planted in the tropics. It was found growing at Trants Estate. *Sight record.*

I observed two Montserratian men making brooms from date palm leaves in a shed at Plymouth.

## ASCLEPIADACEAE

### *Asclepias curassavica* L.; Redhead, Indian Root, Wild Ipecacuanha

This erect, perennial, suffrutescent plant has opposite, lanceolate, acuminate leaves and small orange-red flowers borne in umbels that give rise to lanceolate follicles. It is native to the American tropics. Specimens were collected 1 km west of Galway Soufriere. *Brussell C-13.*

A decoction made by boiling the entire plant is imbibed in small doses as an anthelmintic. A root decoction is drunk in small amounts to treat bad colds, asthma, and gonorrhea. Use of this plant for medicinal purposes is hazardous due to its proven poisonous properties.

**Figure 21.** *Phoenix dactylifera*, date palm at Trants Estate.

*Matelea maritima* (Jacq.) Woodson; Man Tree

This trailing vine has opposite cordate leaves and red flowers that give rise to ovoid-lanceolate follicles covered with corky warts. The flattened, rugose, brown seeds are plumed. It is native to the West Indies and coastal Central America. Specimens were collected in the White River Ghaut. *Brussell C-36.*

Tea made by boiling the leaves is drunk to treat intestinal gas and indigestion.

## ASTERACEAE

*Artemisia dracunculoides* Pursh.; Tarragon

This erect aromatic herb has alternate strap-shaped leaves and many small spherical inflorescences that give rise to obovoid achenes. It is native to the Mediterranean region. Tarragon was found growing in a dooryard garden at Salem. *Brussell C-16.*

Tea made from the leaves is drunk to treat hoarseness.

The fresh leaves are put in salads, and the fresh or dried leaves are used for seasoning purposes. Dried bundles of tarragon are sold in the public market.

*Bidens pilosa* L.; Duppy Needles, Spanish Needles, Beggartick

This annual herb has opposite bipinnate leaves and composite flowers that give rise to linear achenes with barbed awns. It is common in warm regions of the world. *B. pilosa* was collected at Paradise. *Brussell C-183.*

Tea made by boiling the entire plant is sipped to treat colds and to induce abortions. The leaves are boiled, and the resulting decoction is drunk to treat "stoppage of urine." This decoction is said to effectively induce the flow of urine. When Spanish needle leaves are mixed with love vine (*Cassytha filiformis* L.) and boiled, the resulting tea is drunk as a treatment for high blood pressure.

### *Emilia coccinea* (Sims) G. Don; Cabbage Bush

This herb has alternate leaves that vary from pinnately cut to oblanceolate and sagittate. The red flowers are borne on sparse heads and give rise to small brown achenes. *E. coccinea* is native to the Old World but has become widely naturalized in the American tropics. It was collected at the edge of a dooryard garden at St. George's Hill. *Brussell C-319.*

The leaves are cooked and eaten as greens.

### *Emilia sonchifolia* (L.) DC.; Rabbit Food, Cupid's Paint Brush

This herb has alternate oblanceolate or pinnately cut leaves and pink flowers that give rise to very small brown achenes. This Old World native has become naturalized in the American tropics and was found at Woodlands and Cudjoehead. *Brussell C-127, C-333.*

Tea made from the flowers, leaves, and stems is drunk to treat colds. The leaves are fed to rabbits.

### *Erigeron canadensis* L.; Wild Tarragon

This erect columnar herb has alternate linear leaves with numerous small whitish flowers that produce minute achenes. This native of North America has become widely naturalized. It was collected at the edge of a dooryard garden in Woodlands. *Brussell C-136.*

The leaves and upper stems of this plant are used to make a bush tea that is drunk as a general conditioner.

### *Eupatorium odoratum* L.; Christmas Bush

This straggling shrub has opposite ovate leaves and lavender or white flowers that give rise to ribbed achenes. *E. odoratum* is found in the American tropics. It was collected near a footpath leading to the bamboo forest in the Centre Hills. *Brussell C-358.*

A decoction made from the flowers, stems, and leaves is drunk as a treatment for colds and coughs. The crushed leaves are applied topically on cuts as a vulnerary. This plant is fed to cattle after calving to cause the afterbirth to be expelled more rapidly.

### *Eupatorium villosum* Sw.; Sheep Mutton, Sage

This rusty-villose shrub has opposite, ovate, three-nerved leaves and small lavender flowers that produce tiny hairy achenes. *E. villosum* is native to Florida and the West Indies. It was collected at St. George's Hill and at Paradise. *Brussell C-189, C-308.*

Tea made from the leaves is drunk as a cold treatment. The dried leaves are crumbled and used to season stews.

### *Helianthus hirsutus* Raf.; Sage

This erect herb has opposite ovate to lanceolate hirsute leaves and yellow flowers that produce brown achenes. The plant is native to eastern North America and was probably introduced to Montserrat by human activity. It was collected on St. George's Hill. *Brussell C-86, C-313.*

The leaves and stems are used to scour pots and pans.

### *Parthenium hysterophorus* L.; Wormwood, White Broomweed

This erect herb has alternate deeply bipinnately cut leaves and small white flowers that produce tiny, black, broadly obovate achenes. This is a common weed in the American tropics. It was collected on St. George's Hill. *Brussell C-310.*

A decoction made from the leaves, flowers, and stems is drunk as a vermifuge.

### *Pluchea carolinensis* (Jacq.) G. Don; Cure For All

This shrub has alternate, ovate, pubescent leaves and lavender flowers that produce four- or five-angled achenes. *P. carolinensis* is found in the West Indies and from Mexico to Venezuela. It was encountered near a small stream in Woodlands. *Sight record.*

The flowers, leaves, and stems are boiled, and the resulting decoction is drunk to treat stomach ailments and colds.

### *Pseudelephantopus spicatus* (B. Juss. ex Aubl.) C. F. Baker; Cattle Tongue Bush, Bull Tongue

This erect stiff herb has alternate oblanceolate leaves and clusters of white flowers that produce small, pubescent, ribbed achenes. *P. spicatus* is commonly found in the West Indies and continental tropical America. It was collected at Cudjoehead Village. *Brussell C-339.*

A tea made from the leaves is drunk to treat colds, coughs, asthma, and heart trouble.

### *Tanacetum vulgare* L.; Tansy

This erect herb has alternate, oblong, pinnately divided leaves. The yellow showy flowers are borne in dense corymbs and produce angled achenes. This Old World native has become naturalized in diverse habitats around the world. It was found at Tar River Estate and along a road at Plymouth. *Brussell C-70.*

A decoction of the flowers and leaves is imbibed to treat indigestion and female disorders and to induce abortion. A concoction made of equal parts of rum, tansy tea, and chopped green onions is taken in one cup doses four times a day as a vermifuge.

### *Vernonia cinerea* (L.) Less.; Measle Bush

This erect herb has alternate ovate leaves and mauve to purple flowers that give rise to ribbed sub-cylindrical achenes. This is a common weedy species found throughout the tropics. It was collected on Cavalla Hill. *Brussell C-378.*

Tea made from the leaves and stems is drunk to treat measles, coughs, and colds.

### *Wedelia trilobata* (L.) Hitchc.; Pasture Sage, Bobena, Graveyard Grass, Carpet Daisy

This low creeping herb has opposite three-lobed leaves and yellow showy flowers that produce wedge-shaped achenes. This small composite is found from Florida to northern South America. It was collected at Cudjoehead Village and in a pasture at Paradise. *Brussell C-182, C-340.*

Tea made from the leaves, flowers, and stems is drunk to treat asthma and colds as well as to induce abortion.

## BALSAMINACEAE

### *Impatiens balsamina* L.; Balsam

This annual herb has alternate ovate leaves and showy flowers that may be white, pink, red, or purple. The mature capsules suddenly burst open when touched. This native of Asia is widely cultivated and has escaped in many warm areas of the world. It was found at Salem. *Sight record.*

Tea made by boiling the leaves is a pleasant beverage and is said to be a cure for colds (Duberry 1973). It was seen for sale as a bush tea in the Plymouth market.

## BIGNONIACEAE

### *Crescentia cujete* L.; Calabash; plate 5

This small tree has obovate leaves spirally arranged on dwarf shoots and large, greenish-yellow, campanulate flowers that produce small to large globose to oval fruits. The calabash tree is native to tropical America. It was found at Salem. *Brussell C-52.*

Tea made from the flowers is drunk to treat earache. The internal portions of young calabash fruits are boiled, and the resulting decoction is given to victims of asthma attacks and people with cold symptoms. This preparation is said to cause a "rush of blood to the chest."

The inner pulp of mature fruits is, according to some informants, poisonous when consumed in sizable amounts; however, the seeds are cooked and eaten by Montserratians. The gourd-like fruits are used to make vessels for carrying liquids, dippers, bowls, and musical instruments.

*Macfadyena unguis-cati* (L.) A. Gentry; Cat's Claw, Right Wythe

This woody climber has opposite leaves that consist of two lanceolate leaflets and one three-pronged tendril with hooked claws. The large, campanulate, yellow flowers give rise to oblong-linear capsules containing large winged seeds. Cat's claw is native to tropical America and is found throughout the West Indies in mesic forests. It was collected at Cudjoehead Village and Salem. *Brussell C-161, C-343.*

The roots of this plant are soaked in water, and the resulting concoction is drunk as a treatment for a wrenched back. The leaves and stems are boiled to produce a tea that is imbibed to relieve colds.

*Tabebuia pallida* (Lindl.) Miers; White Cedar, Whitewood

This tree has opposite compound leaves with up to five leaflets and small, bell-shaped, yellow flowers that produce long flattened capsules containing winged seeds. White cedar is a common tree of coastal woodlands of the West Indies, Central America, and Venezuela. It was collected at Trants. *Brussell C-198.*

A decoction made from the leaves and twigs is drunk as an antidote for fish poisoning (*ciguatera*).

This is an important timber tree on Montserrat that is used for general construction, furniture, cabinets, interior trims, flooring, and boatbuilding. Both the heartwood and sapwood are light brown, tough, and strong, with a medium to coarse grain. The wood's resistance to splitting is said to be fair. It is easily worked and takes a fine grain finish when sanded.

*Tecoma stans* (L.) Kunth.; Elder Bush, Ginger Thomas, Buttercup

This shrub or tree has opposite compound leaves made up of one to three pairs of lanceolate leaflets and showy, yellow, campanulate flowers that produce elongated beaked capsules. Elderbush is found throughout tropical America. It was collected at Salem. *Brussell C-377.*

Tea made from the leaves, bark, and roots is imbibed to treat colds.

## BIXACEAE

*Bixa orellana* L.; Annatto, Lipstick Plant; plate 6

This shrub or small tree has alternate cordate or ovate leaves and showy pink flowers that produce ovoid red capsules covered with soft bristles. This plant is native to continental tropical America and was probably spread through the West Indies by early Indian migrations. It was found growing on the McChesney Estate at Olveston. *Brussell C-243.*

Tea made from the leaves is drunk for its purgative and diuretic properties. Oil extracted from the seeds is applied to burns.

The arils surrounding the seeds are the source of an important red dye that can be obtained by cooking the seeds in oil or water. This dye is used to color rice, butter, soups, and other foods. Now that carcinogenic properties have been attributed to several synthetic red dyes, it seems that natural coloring agents like annatto will once again be prominent items of commerce. Perhaps Montserrat and other Caribbean Islands could grow and export this product as a cash crop to provide more jobs and stimulate their economies.

## BOMBACACEAE

*Ceiba pentandra* (L.) Gaertn.; Silkcotton; plate 7

This large tree has alternate palmately compound leaves with three to five lanceolate leaflets and showy white flowers that produce large, obovoid, elongated capsules containing many seeds surrounded by dense masses of wool. This tree is native to the West Indies and northern South America. It was collected in a forest above Salem. *Brussell C-176.*

The fluffy "wool" from the fruits is the kapok of commerce and is used locally to stuff pillows, mattresses, and voodoo dolls. When the occupants of a house are gone, one of these voodoo dolls is placed over the front door to keep out intruders and thieves.

The lightweight wood is used for general construction and to make coffins and dugout canoes. During my second sojourn on Montserrat, I witnessed the carving of a dugout canoe from a large silkcotton log. The artisans enlarged the canoe by attaching custom-shaped sideboards to the top edge of the structure.

## BORAGINACEAE

*Bourreria succulenta* Jacq.; Strongback, Strongbark, Pigeonberry

This small tree has alternate lanceolate leaves and large, showy, white flowers that produce yellow drupes. It is native to tropical America and was found at Paradise. *Sight record.*

The root of this plant is soaked overnight in water, and the resulting infusion is drunk to alleviate back pain. It is not clear whether this treatment is meant for the back muscles or for the kidneys.

The hard brown wood is used to make charcoal.

*Cordia alliodora* (Ruiz & Pav.) Oken; Seepwood, Cypre

This tree has alternate oblong leaves and small white flowers that produce ellipsoid drupes. Seepwood is native to the American tropics and was found in a mesic woods above Salem. *Brussell C-162.*

Tea made from the seeds and leaves is drunk to relieve colds.

The pale light-brown wood is highly valued for furniture, cabinet work, and general construction.

*Cordia collococca* L.; Clammy Cherry, Red Manjack

This tree has alternate ovate leaves and white flowers that produce scarlet globose drupes. It is native to tropical America and was found near the Yacht Club. *Sight record.*

The sticky pulp of the fruit is used as glue. Playful children throw the fruits at one another and take delight in the fact that the crushed fruits will stick to a person's back for quite some time without that person knowing it.

*Cordia curassavica* (Jacq.) **Roem. & Schult.**; Cow's Tongue, Saize Bush, Sage Bush, Langford Cabbage

This shrub has aromatic, alternate, ovate to lanceolate leaves that are hirsute on the upper surface and white flowers that give rise to small red globose drupes. It is native to tropical America and was collected in the White River Ghaut and in a thorn forest at Olveston. *Brussell C-43, C-94.*

Montserratians use the leaves as an ingredient in bush tea that is drunk as a beverage and as a cold medicine. A decoction made from the leaves and twigs is drunk to reduce high blood pressure. "Young sage bush leaves and young sugar cane leaves" are put into a covered jar of sea water and allowed to remain for three or four days. The resulting concoction is drunk in one cup doses four times per day for five days to "get rid of gonorrhea."

The coarse, scabrous leaves are used to scour pots and pans.

*Cordia nitida* **Vahl**; Leley, Red Manjack

This small tree has whorled elliptic leaves, clusters of white campanulate flowers, and scarlet globose drupes. It is found in the West Indies and Central America and was collected in the White River Valley. *Brussell C-104.*

Some Montserratians regard the slightly astringent fruits as being poisonous, while other people of the island eat the fruits raw or cooked in pies.

*Cordia obliqua* **Willd.**; Clammy Cherry, White Manjack; figure 22

This small tree has alternate, broadly ovate, elliptic, or suborbicular leaves and yellowish-white flowers that produce pale yellow subglobose drupes. It is native to India and was collected at the Yacht Club Beach. *Brussell C-1.*

I observed children chewing on the slightly sweet, juicy, mucilaginous, fruits. The sticky pulp of the fruits is used as glue.

*Tournefortia filiflora* **Griseb.**; Elder

This erect suffruticose plant has large, alternate, elliptical-oblong leaves. The flowers, which are borne in corymbose-panicled divaricate spikes, have corolla tubes four to five times as long as the calyces and give rise to

Figure 22. *Cordia obliqua*, clammy cherry.

subglobose glabrous drupes. It is native to the West Indies and has heretofore not been reported from Montserrat. Specimens were collected above Salem. *Brussell C-174*.

The leaves are used to make poultices for treating wounds. A decoction made by boiling the mature fruits is used for medicinal purposes.

*Tournefortia volubilis* L.; Sea Lavender, Chigger Nut, Soldier Bush

This liana has alternate lanceolate leaves and small, cylindrical, greenish-yellow flowers that give rise to white drupes with four black spots. This vine is distributed from Florida to South America and throughout the West Indies. It was found at Woodlands Beach. *Sight record*.

An insecticidal powder is made from the ground bark and leaves.

A decoction of the leaves and roots is used as a hair and body cleanser. The leaves are used as poultices to treat blisters and wounds.

## BROMELIACEAE

*Ananas comosus* (L.) Merr.; Pineapple, Pine

This coarse herb has spirally arranged linear leaves with saw-toothed margins. The inflorescence is a dense flat-topped head surrounded by a leaf rosette. The large, sweet, succulent compound (syncarpous)

fruit is topped with a rosette of leaves. It is native to northern South America and was found in many dooryard gardens on Montserrat. *Sight record.*

Slices of the fresh ripe fruit are placed on wasp stings as a poultice to relieve pain and break down the venom. Sea urchin wounds are first treated with a topical application of fresh urine, followed by repeated fresh pineapple-fruit poultices. The pineapple poultices are used for about two hours, after which the affected area is washed and dried and allowed to heal, usually without a bandage. Fresh pineapple slices are eaten to relieve an upset stomach.

Pineapple fruit is commonly eaten raw and cooked in various cakes, puddings, and other dishes.

## BURSERACEAE

***Bursera simaruba*** (L.) Sarg.; Gum Tree, Gummy Lingo, Gum Bark, Gum Bush, Cum Bush; figure 23

This tree has alternate pinnately compound leaves with lanceolate leaflets and small, unisexual, yellowish-green flowers that give rise to small three-angled drupes. This tree is found from southern Florida to northern South America. *B. simaruba* was collected in the Centre Hills and in the White River Ghaut. *Brussell C-39, C-41, C-87.*

A decoction of the bark is drunk to "purify the blood," to reduce fevers, ease stomach distress, and

Figure 23. *Bursera simaruba*, gum tree, with *Phoradendron trinervium*, no mammy, growing on it.

relieve congestion of the sinuses and lungs. A leaf decoction is drunk as a treatment for coughs and colds.

The gum that exudes from the trunk and limbs is burned as an aromatic incense and has been used in voodoo practices on the island.

The trunks of gum trees are set out as living fence posts, due to the tendency of these fresh cut poles to form roots and leaves and begin to grow.

## CACTACEAE

*Cephalocereus royenii* (L.) Britton & Rose; Dildo, Pipe Organ Cactus; figure 24

This tree cactus has erect columnar branches, leaves reduced to spines, large greenish-pink showy flowers, and globose red berries. It is native to the West Indies and was found near Shoe Rock. *Sight record.*

The sweet juicy fruits are edible raw or cooked into jam or syrup. The taste is suggestive of a strawberry-watermelon combination. The succulent water-storing tissue inside the trunk can be used as a source of sterile water. The flesh from this cactus is dried, pulverized, and used as a base for a local soup.

*Hylocereus trigonus* (Haw.) Safford; Nightblooming Cereus, Wild Strawberry

This slender climbing cactus has leaves that are reduced to spines, very large nocturnal white flowers,

**Figure 24.** *Cephalocereus royenii*, dildo.

and large red oblong-obovoid berries. It is native to Puerto Rico, Hispaniola, and the Lesser Antilles. Nightblooming cereus was collected at Woodlands. *Brussell C-199.*

The fruits are eaten fresh for their laxative effect.

Aside from their use as a laxative, the fruits are eaten raw for enjoyment and cooked in preserves.

*Melocactus intortus* (Miller) Urban ( = *Cactus intortus* Miller); Turk's Cap, Turk's Head; figure 25

This depressed-globose succulent has leaves that are reduced to stout subulate spines, large pink flowers, and lavender-pink obovoid berries that contain many tiny seeds. This cactus is known only from the West Indies and was found near Shoe Rock. *Sight record.*

The shiny, lavender-pink, succulent fruits are edible raw and have a flavor that resembles a strawberry-watermelon combination. The pulp of this cactus is a source of water in survival situations.

*Opuntia cochenillifera* (L.) Mill.; Cochineal Cactus; figure 26

This spineless succulent has a cylindrical trunk, flattened obovate branches, small subulate leaves, and crimson flowers that produce red obovoid berries. This cactus is native to Mexico but has been widely cultivated in the American tropics. It was found near the Radio Antilles building. *Sight record.*

Figure 25. *Melocactus intortus*, Turk's cap.

**Figure 26.** *Opuntia cochenillifera,* cochineal cactus.

The flat stems of this plant are peeled and used as poultices to treat swelling and pain. These flat "pads" are also sliced open, and the exuding juice is used by some Montserratians to wash their hair.

This plant was once cultivated on Montserrat as a host for raising cochineal insects. These scarlet-colored insects were once an important source of red dye.

### *Opuntia dillenii* (Ker-Gawl.) Haw.; Prickle Pear

This erect spine-bearing succulent has jointed branching stems and small, deciduous, sublate leaves. The large, showy, yellow flowers give rise to purplish-red pyriform fruits. This cactus occurs in Florida, Bermuda, the West Indies, and the eastern coast of Mexico. It was found near the Radio Antilles building. *Sight record.*

The flat stems are seared in a fire to remove the sharp spines and split in half longitudinally. The moist inner surface is then placed on wounds and swollen injuries as a poultice. This poultice is placed on sea urchin punctures for one day, after which the spines can easily be removed. The fruit is also used in the same manner as a poultice.

Cattle herders use the slimy juice of this plant to "bring back the cud to cattle" when they are having difficulty regurgitating their food for a second chewing. The juice is administered to the animal orally with a wine bottle.

Montserratians enjoy eating prickle pear fruits

fresh and cooked in pies and jam. Before eating these fruits, one must be certain to burn or peel off all of the spiny glochids.

The pads are cut in half, and the slimy juice is mixed with water and put on the hair as a conditioner that is said to beautify and stimulate growth of the hair.

## CAMPANULACEAE

*Hippobroma longiflora* (L.) G. Don; Star of Bethlehem

This erect perennial herb has alternate oblanceolate leaves with coarsely wavy-dentate margins. The white flowers give rise to nodding ovoid capsules. It is native to tropical America. Specimens were collected in the Galway Soufriere area. *Brussell C-26.*

This plant is very toxic and has been employed as a poisoning agent. The sap can cause blindness.

## CANNACEAE

*Canna edulis* Ker-Gawl.; Toiseloisemoise, Toloma Food, Mountain Porridge; figure 27

This tall herb has fleshy tuberous rhizomes, ovate leaves contracted into the sheath, and large, showy, red flowers that give rise to warty capsules with many subglobose black seeds. It is native to Peru and Brazil but has been widely planted in the American tropics. *C.*

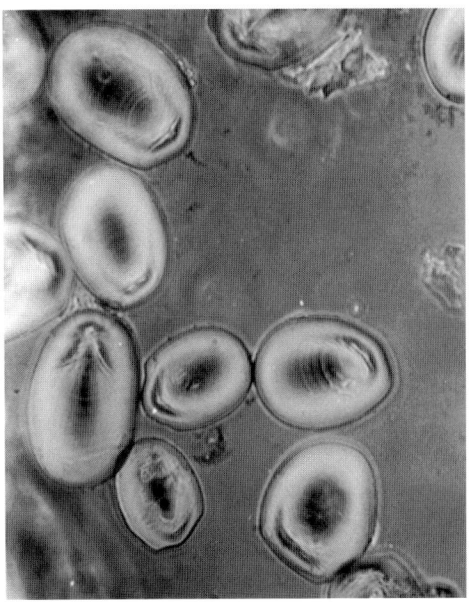

**Figure 27.** Phase contrast micrograph (250×) of *Canna edulis*, toloma, starch grains. (*Brussell C-188*). Small fragments of archaeological remains of the rhizomes of *C. edulis* have been identified by comparing starch grains from ancient material to starch grains of modern specimens. Photo courtesy of Dr. Donald Ugent.

*edulis* was collected in an upper level garden at Paradise. *Brussell C-188.*

The starchy tuberous rhizomes are baked and eaten like potatoes or boiled and made into porridge that is said to be "fattening."

Archaeological and linguistic evidence point to cultivation of *C. edulis* in Peru perhaps as early as 8000 B.C. Identification of archaeological *C. edulis* rhizomes has been accomplished by comparing the starch grains of ancient material to the starch grains of modern specimens (Ugent et al. 1984).

## CAPRIFOLIACEAE

*Sambucus simpsonii* Rehd.; Elder

This small tree has opposite imparipinnate to bipinnate leaves and large sprays of small white flowers that produce small reddish-purple drupes. It is native to the southeastern United States and was collected at Woodlands. *Brussell C-138.*

The leaves and flowers are used to make a relaxing tea, and the fruits are used to make syrup and wine.

## CARICACEAE

*Carica papaya* L.; Papaya, Pawpaw

This small tree or treelike herb has alternate deeply lobed leaves and unisexual or hermaphroditic, showy, ivory-white to yellow flowers that give rise to large, fleshy, yellow fruits containing many rough seeds. It is native to tropical America and was found at Trants. *Sight record.*

A tea made from the leaves is taken in one cup doses three times a day for kidney trouble. Sap from the leaves and stems is applied externally to treat ringworm. The green fruit is grated and mixed with milk, and the mixture is then drunk to quickly reduce blood pressure. The unripe green fruits are sliced open and placed on wasp and bee stings to relieve pain and reduce swelling. I had occasion to use the fruit in this manner after being stung on the leg by three wasps while on a collecting trip. It proved to be an effective remedy. It is speculated that the proteolytic enzyme (papain) in the fruit may denature wasp venom. The green fruit is also used as a poultice to help wounds heal. The seeds are chewed to allay indigestion, and the green or ripe fruits may be eaten for the same purpose.

The ripe fruit is highly prized when eaten fresh and the juice of this fruit is also popular on the island. Green papayas are pickled or candied and used in confections and chutneys. The green fruit can be used to make a tasty casserole: It is peeled, halved, and cut into small pieces. The pieces of fruit are cooked in sea water or salted fresh water until tender and drained. The drained squares of papaya are then placed in a buttered dish and sprinkled with black pepper, grated cheese, and fresh grated ginger. Creamed corn is then spread over the top, and the mixture is baked until golden brown.

The seeds are sometimes ground and used as a substitute for black pepper.

## CHENOPODIACEAE

*Chenopodium ambrosioides* L.; Worm Oil Tree, Wormwood, Wormseed Weed

This aromatic herb has alternate narrowly elliptic leaves and minute green flowers that produce small globular utricles. This plant is widely distributed in warm areas of the world and was collected west of Galway Soufriere, at Salem, and St. George's Hill. *Brussell C-7, C-98, C-314.*

A decoction of the leaves is drunk as an anthelmintic.

## CLUSIACEAE

*Clusia rosea* Jacq.; Wild Mammy Apple, Wild Mansipote, Scotch Attorney, Coretor

This tree has opposite, obovate, thick, rigid leaves and white or pink flowers that give rise to globose leathery capsules. It is native to the West Indies and continental tropical America. *C. rosea* was collected at Galway Soufriere and on a hill above Salem. *Brussell C-19, C-156.*

The yellowish latex from the bark and fruit is applied topically to get rid of scar tissue. It hardens after being exposed to the air and is used to caulk boats and as a glue.

*Mammea americana* L.; Mammy, Mammy Apple; plate 8

This tree has opposite obovate leaves and white flowers that produce large reddish-brown drupes with yellow flesh and large seeds. It is native to the West Indies and continental tropical America. *M. americana* was collected at Woodlands. *Brussell C-14, C-85, C-281.*

Tea made from the leaves is drunk to reduce fevers.

The large fruits are eaten raw or cooked in pies and other dishes. The leathery rind, which is bitter due to its tannin content, must be peeled away before the firm yellow flesh can be eaten. The fresh fruit is sweet, and mammy apple pie is similar to peach cobbler.

The seeds are ground, and the resulting powder is used as an insecticide on pets and garden plants. The powder made from the seeds is mixed with water and applied externally to kill fleas on dogs. The resin from the bark is used topically to extract chiggers and to kill ticks.

## COMBRETACEAE

*Terminalia catappa* L.; Beach Almond, Indian Almond

This tree has alternate obovate leaves and small greenish-white flowers that produce yellow flattened-ellipsoid drupes with thin flesh. It is native to the East Indies and Oceania, but has been widely naturalized on tropical shores around the world. The buoyant corky tissue in the pericarp of the fruits causes them to float and facilitates marine seed dispersal. *T. catappa* was collected at Lime Kiln Beach. *Brussell C-302*.

The seed kernels are eaten like nuts. The outer flesh has a sweet-tart fruity flavor and can be eaten fresh. The hard strong wood is used for general construction, boatbuilding, flooring, and bridge timbers.

## COMMELINACEAE

*Commelina diffusa* Burm.; White Frenchweed

This herb has alternate ovate-lanceolate leaves and blue flowers that give rise to three-celled capsules, each with two reticulate seeds. This pantropical plant extends partially into the temperate regions. It was collected at Olveston. *Brussell C-97*.

Tea made from the leaves is drunk to induce urine flow in old men.

*Commelina elegans* Kunth; Water Grass, White Weed

This herb has alternate lanceolate leaves and blue flowers that produce small capsules, each containing three smooth seeds. It is native to tropical America. Specimens were collected at Cudjoehead. *Brussell C-342*.

The succulent leaves and stems of this plant are pulverized into a mucilaginous paste that is applied to cuts and other wounds to stop bleeding.

Tea made from the leaves is drunk to induce urine flow in old men and to treat colds and inflammation.

## CONVOLVULACEAE

*Ipomoea batatas* (L.) Lam.; Sweet Potato

This twining herb has alternate cordate leaves and showy white to pink flowers that produce ovoid capsules containing dark glabrous seeds. It is native to tropical America. Sweet potatoes were found in a dooryard garden at Woodlands. *Sight record*.

The whole "potato" is boiled and the resulting decoction is drunk for reducing high blood pressure. Sweet potato slices are tied onto both jaws to treat mumps (Duberry 1973).

Sweet potatoes are a common food source on Montserrat and are sold in the public market.

*Ipomoea pes-caprae* subsp. *brasiliensis* (L.) Ooststr.; Beach Morning Glory; plate 9

This glabrous creeping herb has alternate oblong to roundish leaves with a notched apex and subcordate base, pink to purple showy flowers, subglobose two-celled capsules, and reddish-brown pubescent seeds. The beach morning glory is widespread on tropical beaches. It was collected at the Yacht Club Beach. *Brussell C-2.*

The seeds, which are said to have hallucinogenic properties, are known to be poisonous. They have been used in voodoo practices on the island.

*Merremia umbellata* (L.) H. Hallier; Hog Vine

This twining vine has alternate long-cordate leaves and large yellow flowers that give rise to subglobose, four-valved, four-seeded capsules. It is found in the West Indies, Central America, and the Old World tropics. *M. umbellata* was collected in a weedy dooryard garden at Woodlands. *Brussell C-76.*

The underground portions of this plant are washed, macerated, and consumed fresh by a mother after parturition. Tea made from the underground portions is drunk to treat jaundice and dysmenorrhea.

*Turbina corymbosa* (L.) Raf.; Christmas Bush, Christmas Wreath

This climbing vine has alternate cordate leaves with large white flowers that give rise to ellipsoid capsules that each contain one pubescent seed. *T. corymbosa* is distributed in the West Indies and continental tropical America. It was collected on a field border at Paradise. *Brussell C-192.*

The entire plant is boiled and given to cows to expel afterbirth following parturition.

## CRASSULACEAE

*Bryophyllum pinnatum* (Lam.) Oken ( = *Kalanchoe pinnatum* [Lam.] Pers.); Love, Love Bush, Love Leaf, Leaf of Life

This erect succulent has opposite lower leaves that are simple and lanceolate. The upper leaves are pinnate with lanceolate leaflets. The leaves produce young plantlets on the margins. The showy red flowers produce green fruits consisting of four oblong carpels. This native of Madagascar is now widespread in the tropics. It was found in a roadside ditch in Woodlands. *Brussell C-146.*

The leaves are boiled and the resulting decoction is drunk to treat colds, sluggish kidneys, and high blood pressure. The tea made from the flowers is said to have mild opiatelike effects. The juice squeezed from the leaves and stems is mixed with salt and swallowed to get rid of phlegm. The leaves are macerated and mixed with sweet oil (vegetable oil) to make a poultice that is used to reduce swelling.

## CUCURBITACEAE

*Cucumis anguria* L.; Wild Cucumber, West Indian Gherkin

This trailing herb has alternate deeply five-lobed leaves and small yellow flowers that produce pale-yellow, prickly, ovoid, fleshy berries. This plant is found in moist soil in the West Indies and continental tropical America. It was collected in an old field near the Vue Pointe Hotel. *Brussell C-326.*

The fruits are eaten like common cucumbers.

*Luffa aegyptiaca* Mill.; Loofah

This climbing twiner has tendrils, alternate five-lobed leaves, and yellow flowers that produce a fruit the size of a cucumber. This Old World native is widely planted in tropical and subtropical areas of the world. It was found at Salem. *Sight record.*

The fibrous, peeled "skeleton" of the fruit is used for bathing and scrubbing, as well as for straining liquids. The dried fruits were seen in the public market.

*Momordica charantia* L.; Maiden Apple, Pom Cooly, Pung Cooly, Lizard's Food; plate 10

This slender vine has simple tendrils, alternate deeply palmate-lobed leaves, and unisexual yellow flowers that produce orange, tuberculate, ovoid fruits that contain numerous seeds covered with red arils. It is native to the Old World tropics. Specimens were collected at Cudjoehead and Plymouth. *Brussell C-3, C-303, C-344.*

The bitter leaves of this plant are salted, pounded, and boiled in water to make a decoction that is taken to cure a very bad cold or sore throat and as a purgative. Strong doses are said to cause abortion.

A decoction made by boiling the young leaves in water is drunk to relieve lung congestion and asthma. Tea made from the mature leaves is sipped as an influenza remedy and as an agent for "cooling the blood" (reducing high blood pressure).

A Jamaican living on Montserrat claims she cured herself of stomach cancer by drinking tea made by boiling the crushed seeds. She drank this tea in one-cup doses four times a day for three months.

A decoction made from the leaves and stems is mixed with lime juice and drunk in one-cup doses four times a day as a treatment for sugar diabetes. It is said to be quite effective.

The sweet red arils are eaten for indigestion, fevers, and heart trouble.

Wrapping the vines around the neck of a person with a high temperature is said to get rid of the fever, according to a Montserratian voodoo legend. The vines are also used to stop pain. The ailing body part is wrapped with large quantities of the vine which are left on for one night to relieve the pain. It is said that this treatment will only work if the vine is gathered prior to six o'clock in the evening (Duberry 1973).

An antiviral protein from this plant shows potential as a therapeutic agent for treating AIDS (Lee-Huang et al. 1990).

*Sechium edule* (Jacq.) Sw.; Christophine, Chayote; plate 11

This herbaceous vine has tendrils, alternate cordate five-angular leaves, and yellow flowers that produce obovoid, soft-prickled, green fruits each with one large seed. It is native to the West Indies. Specimens were collected in a dooryard garden at Woodlands. *Brussell C-144.*

Tea made from the fruits and leaves is drunk to lower blood pressure.

The fruits are cooked as vegetables and are also pickled. The young shoots and tuberous roots are also eaten cooked.

## DIOSCOREACEAE

*Dioscorea alata* L.; Yam

This high-climbing vine has opposite cordate leaves, small male flowers, larger female flowers in simple spikes, and elliptic three-winged capsules. It is native to southeastern Asia but is widely cultivated in warm areas of the world. *D. alata* was found in a dooryard garden at Harris Village. *Sight record.*

The yam serves as an important food crop on Montserrat; however, the immature yams should not be eaten due to the presence of oxalates and cyanogenic glycosides, which are poisonous (Fritzmaurice 1953).

## EBENACEAE

*Diospyros revoluta* Poir.; Black Apple; plate 12

This tree has alternate ovate leaves and dioecious, small, white flowers that give rise to globose black berries. It is native to the West Indies. Specimens were collected at Salem and Woodlands. *Brussell C-131, C-240.*

The chipped bark and crushed seeds of this tree are put in a basket and dragged through the water in a circle. Active principles leached from the bark and seeds "burn the eyes" of fish and stun or kill them. Swimmers then go after the fish.

The heartwood is black and very hard.

One informant reported that the fruits are edible.

## EUPHORBIACEAE

*Adelia ricinella* L.; Batroot

This shrub has alternate obovate leaves and clusters of pubescent white flowers that give rise to pubescent capsules. It is native to the West Indies. Specimens were collected at Woodlands and in the White River Valley. *Brussell C-54, C-113.*

All parts of the plant are boiled, and the resulting decoction is drunk for fevers and body pain.

***Croton flavens* L. (= *Croton balsamiferum* Jacq.); Balsam, Sweet Balsam, Rock Balsam, Yellow Balsam, Seaside Sage**

This aromatic shrub has alternate ovate leaves that are tomentose on both surfaces. The small, white, fragrant flowers are borne in terminal racemes and give rise to small subglobose capsules, each containing three black seeds. *C. flavens* is native to the West Indies. It was found in the White River Valley, in the White River Ghaut, at Bugby Hole, and Shoe Rock. *Brussell C-37, C-99, C-204, C-261.*

Tea made from the leaves is drunk as a stimulating beverage and as a cold medicine. A decoction made from the leaves and branches is used to wash wounds and is applied as a bath for skin rashes. The macerated fresh leaves are employed topically in poultices to treat burns. The dried leaves are smoked to treat lung congestion and as a tobacco substitute.

***Croton flocculosus* Geisl.; Bitter Balsam**

This shrub has alternate, ovate-lanceolate, long-petioled leaves and small white flowers in terminal racemes that give rise to globose capsules. It is native to the West Indies. Specimens were collected in a rocky ghaut at Olveston. *Brussell C-348.*

Tea made from the leaves is drunk for asthma. The dried leaves are smoked to treat lung ailments.

***Euphorbia heterophylla* L.; Jacob's Ladder, Red Milkweed**

This erect suffrutescent herb has alternate leaves that vary from orbicular through various shapes to linear, with green male and female flowers in a cyathium and capsules bearing small, ovoid, pointed, gray seeds. It is distributed from Illinois to Montana and southward to South America and the West Indies. It has been introduced into the Old World. On Montserrat it was collected on a hillside at Woodlands. *Brussell C-126.*

This plant is said to be poisonous to rabbits and has been employed as a poisoning agent.

***Euphorbia maculata* L. var. *thymifolia* L.; Eyebright**

This cespitose annual herb has opposite oblong-rectangular to broadly elliptic-ovate leaves and brownish-white to pinkish-green flowers that give rise to hairy capsules containing brownish tetragonous seeds. It is native to the American tropics and subtropics. Specimens were collected at Trants Estate. *Brussell C-201.*

Juice squeezed from the leaves and stems is diluted with water and used in small amounts as eyedrops to "clear the eyes." Diluted tea made by boiling the leaves is imbibed to treat stomach ailments. Overdoses of these treatments can be toxic.

***Euphorbia pulcherrima* Willd.; Poinsettia, Christmas Flower**

This shrub has alternate ovate-elliptic leaves and orange and green flowers subtended by large, showy, red or white bracts. The three-lobed fruit has three oblong seeds. This plant is native to southern Mexico. It was found at Olveston. *Sight record.*

As is the case with other species of *Euphorbia*, the sap is caustic. The sap is used as a depilatory agent after being diluted with sweet oil in order to diminish its corrosive properties. It should be noted that the sap is damaging to the eyes. Some people who come into contact with this plant (especially the sap) develop a skin rash that is accompanied by inflammation of the skin and by itching and burning sensations.

***Euphorbia tirucalli* L.; Pencil Plant, Rubber Hedge Plant**

This shrub has a great many rubbery branches, small deciduous alternate oblong leaves, and minute yellow flowers. It is native to Rhodesia and has been planted in many tropical areas of the world. On Montserrat it was collected at Lime Kiln Beach. *Brussell C-154.*

The caustic, poisonous, white sap is used as a rat poison and as an insecticide.

***Hippomane mancinella* L.; Manchineel**

This evergreen tree has alternate elliptic leaves with purple male flowers and red female flowers that give rise to yellow sweet smelling drupes that contain smooth flattened seeds. It is native to tropical America. Specimens were collected at Vue Pointe Beach. Ocean currents distribute the seeds. *Brussell C-150.*

Small amounts of the caustic poisonous sap are used externally with great caution to remove warts. Amerindians poisoned their arrows with the sap.

The manchineel is said to be the most poisonous tree on Montserrat. The greenish-yellow fruits are sweet-scented and agreeable in taste but very toxic. Ship-wrecked sailors, conquistadores, and modern-day tourists have died from ingesting the fruits.

Hogs have been poisoned as a result of eating the fruits, and cattle have been afflicted with skin irritation from rubbing against the trees. On Montserrat, it is generally believed that a person who sits under a manchineel tree during a rainstorm will receive blisters from the water dripping off the leaves and stems.

***Hura crepitans* L.; Sandbox Tree; plate 13**

This large tree has alternate cordate leaves with red male flowers and pinkish-yellow female flowers that produce large capsules that dehisce explosively in dry weather, thus scattering the seeds. The sandbox tree is native to the West Indies. It was found at Olveston. *Brussell C-357.*

The sap causes dermatitis; however, small amounts of it are used with great caution externally to remove warts. The poisonous sap is also used to stupefy fish. The seeds are used as rat poison.

People and animals are sometimes frightened and/or injured when the seed capsules of this tree explode.

***Jatropha curcas*** L.; Body Cutter

This small tree or shrub has alternate roundish, angular, or three- to five-lobed leaves with yellow flowers arranged in corymbose cymes that give rise to ellipsoid, slightly fleshy capsules each containing three oblong peanut-like seeds. It is native to the American tropics. Specimens were collected in the White River Ghaut. *Brussell C-46.*

A decoction made from the leaves is applied topically to stop bleeding. Fresh or dried leaves are ground up, mixed with petroleum jelly and put on cuts or sore gums.

Eating the seeds causes gastroenteritis.

***Jatropha gossypifolia*** L.; French Body Cutter

This single-stemmed shrub has alternate, three- to five-partite, cordate leaves and deep purple flowers that produce three-furrowed, truncated capsules that contain mottled oily seeds. This plant is native to the West Indies. It was collected at Shoe Rock and White River Ghaut. *Brussell C-49, C-260.*

The entire young plant is boiled to make a decoction that is drunk to treat high blood pressure and diabetes. This same decoction is used externally as a bath for skin rashes. The fresh leaves are soaked in sweet oil and used as suppositories for hemorrhoids. A poultice made from the crushed fresh leaves is applied to cuts, wounds, and tumors.

***Manihot esculenta*** Crantz; Cassava; figures 28, 29, 30, 31, and 32

This cultivated shrub has large alternate semicircular leaves that are divided into several lanceolate lobes. The purple flowers give rise to oval capsules that contain mottled brown seeds. It is native to Brazil. Specimens were collected at Salem. *Brussell C-89, C-92.*

The fresh bark is applied to wounds as a healing bandage.

There are bitter and sweet forms of cassava that provide an important dietary staple. The starchy roots of either may contain poisonous cyanogenic glycosides which must be removed before the roots can be utilized as a food source. The roots are peeled and ground or pounded to release the glycosides and enzymes. The wet pulp is left to sit for twelve to twenty-four hours, during which time a portion of the volatile hydrogen cyanide gas escapes. Thorough cooking expels the remaining

Figure 28. *Manihot esculenta,* cassava, the source of tapioca and cassava meal.

Figure 29. Informant James Cabey grinding cassava.

**Figure 30.** Informants displaying dried cassava meal, with toxic compounds removed, that will be used to bake cassava bread, a dietary staple.

**Figure 31.** One of the few remaining eighteenth-century stone ovens used for baking cassava bread. The long wooden forks are used to turn the bread.

**Figure 32.** Cassava bread drying.

poisonous gas. Cassava bread can be baked directly from the moist pulp or the wet grindings can be dried, powdered, and stored for later use. A finer grade of starch can be recovered from the juice extracted from the wet pulp by decanting the liquid after the starch has precipitated out. When the starch is heated in a metal pan, it forms gelatinous granules called tapioca. The starchy internal region of the roots of sweet cassava contains very little or no poisonous glycosides. In Africa sweet cassava roots are simply peeled and boiled. Boiling breaks down the scant cyanogenic glycosides in the peeled roots and expels the hydrogen cyanide gas. The bark of the roots of sweet cassava contains dangerous amounts of cyanide compounds and poisoning could result from eating cooked sweet cassava with the bark left on. The peeled boiled roots of sweet cassava are utilized to produce various cassava products (Simpson and Conner-Ogorzaly 1986).

On Montserrat, bread, porridge, pudding, and coconut sweet (a confection) are made from fresh or dried cassava meal. I enjoyed eating cassava bread baked in a charcoal-fired eighteenth-century stone oven. The taste was similar to white wheat bread toast with a slight nutty flavor.

### *Phyllanthus acidus* (L.) Skeels; Gooseberry Tree

This small tree has alternate pinnate leaves with ovate leaflets. The minute pink flowers produce yellow, juicy, sour berries. It is native to tropical Asia and specimens were collected at the McChesney Estate in Olveston. *Brussell C-250.*

A decoction made from the roots and seeds is applied externally to insect bites and sprains to relieve pain.

The fruits are eaten raw or used to make wine, jam, pies, raisins, and pickles.

### *Phyllanthus anderssonii* Muell.; Red Iron Bark

This small shrub has alternate ovate leaves, small ovate male flowers confined to the lower axils, and saucer-shaped female flowers borne in terminal racemes and axillary clusters. The small, flattened, three-furrowed capsules contain three-sided seeds. Gooding et al. (1965) states that this plant is endemic to Barbados; however, now it is also known from Montserrat. It was collected on a mesic hill above Salem. *Brussell C-169.*

The bark is boiled, and the resulting decoction is drunk to treat high blood pressure.

### *Ricinus communis* L.; Castor Nut, White Castor Nut, Oil Nut, Castor Oil Plant

This large semi-woody shrub has alternate, peltate, nearly orbicular leaves and numerous small, green, apetalous flowers that produce ellipsoid spiny capsules containing mottled, brown to black, flattish-ellipsoid seeds. This Old World native is widely cultivated around the world. It was found at Woodlands. *Brussell C-83*.

Tea made from the leaves is drunk to treat high blood pressure, skin rashes, stomach distress, and fevers. The fresh leaves are heated and used as poultices for treating pain and swelling. The seeds contain a poisonous principle which causes severe gastroenteritis.

## *Sapium caribaeum* Urban; Poison Tree, Bird Lime

This evergreen tree has alternate elliptic-oblong to lanceolate-oblong leaves and spikes of many minute flowers that give rise to orbicular purple capsules that each bear two red seeds. Abundant milky sap is produced by this tree. *S. caribaeum* is endemic to the Lesser Antilles. It was found in a mesic forest at Paradise. *Sight record*.

The caustic latex causes blisters on the skin of a person who comes in contact with it. It is used as a poisoning agent and as an insecticide.

## *Tragia volubilis* L.; Cow Itch, Vine Nettle

This twining, slightly woody, slender vine has alternate cordate leaves, small inconspicuous green flowers, and small three-lobed hispid capsules. It is found in tropical America and West Africa. On Montserrat it was collected at Woodlands. *Brussell C-110*.

The leaves and stems are covered with stinging hairs that cause dermal inflammation and itching when touched.

## FABACEAE

### *Abrus precatorius* L.; Jumbie Bead, Crab's Eyes, Wild Licorice, John Crow Beads

This shrubby, semi-woody, high-climbing twiner has alternate abruptly pinnate leaves, pink flowers, and oblong pods containing globose red seeds with a black dot. This native of India has been naturalized in many tropical areas of the world. It was collected at White River Ghaut, Cudjoehead, and Salem. *Brussell C-38, C-236, C-334*.

The seeds, which are poisonous when raw, are crushed and boiled to make a beverage like coffee that is drunk to treat ulcers and to expel intestinal worms. Tea made from the leaves is drunk as a cold medicine and applied externally as a pain-relieving lotion. A decoction made from the roots is applied externally for pain. Powder made by grinding the seeds is applied externally to skin ulcerations and diseased areas of the skin and scalp. This could be dangerous if the skin of the scalp or other body part is broken, allowing entry of the toxic compound, abrin.

In the past, Jumbie seeds were put in the bottoms of kerosene lamps to raise the level of kerosene when the oil was running low (Duberry 1973). The seeds are also used for making necklaces, bracelets, and other ornaments, and as rattling agents in musical instruments.

### *Acacia farnesiana* (L.) Willd.; Sweet Acacia, Friendego; figures 33 and 34

This shrub or small tree has straight stipular spines, alternate bipinnately compound leaves, yellow flowers in globose heads, and swollen pods marked with longitudinal lines. This plant is native to the American tropics. It was found at Woodlands. *Sight record*.

The leaves and branches are boiled, and the resulting decoction is taken internally to relieve cold symptoms. The washed roots are pounded and boiled to make a tea that is imbibed to treat fish poisoning (ciguatera) that results from eating fish that contain toxin. This tea is said to be poisonous in an overdose.

### *Acacia macracantha* Humb. & Bonpl.; Cusha, Casha

This shrub has branches armed with stipular spines, alternate bipinnately compound leaves, yellow flowers in globose heads, and flattened net-veined pods that are minutely tomentose. It is native to the West Indies. A specimen was collected at the Yacht Club Beach. *Brussell C-4*.

The sap and the stem exudate are used to reduce "proud flesh" (scar tissue). Fresh cut stem sections about 12 cm in length are heated in a fire at one end

**Figure 33.** Informant cutting *Acacia farnesiana* for charcoal production.

Figure 34. Charcoal pit filled with wood of *Acacia farnesiana* prior to firing.

until the sap and other plant juices drip out the other end of the stem. The resulting liquid is applied to healing cuts to reduce scar tissue formation and also on old scars to reduce the amount of "proud flesh."

The wood is used to make charcoal and tool handles.

### *Acacia tortuosa* (L.) Willd.; Friendego, Cashaw

This low tree has long stipular spines, alternate bipinnately compound leaves, small yellow flowers in globose axillary heads, and elongated subcylindrical pods. It is native to the West Indies. A specimen was collected at Salem. *Brussell C-159.*

The washed crushed roots are put in a container with water, and the mixture is allowed to ferment. The fermented liquid is then drunk to treat fish poisoning (ciguatera). This concoction is said to be poisonous in overdoses.

### *Adenanthera pavonina* L.; Jumbie Bead Tree

This tree has alternate bipinnately compound leaves, clusters of pale yellow flowers, and brown elongated pods that split and twist open to reveal shiny, scarlet, lens-shaped seeds. It is native to tropical Asia. Specimens were collected at Salem. *Brussell C-294.*

The bright scarlet poisonous seeds are used to make necklaces and for decoration on baskets. The seeds are also associated with voodoo customs. The word "Jumbie" refers to a voodoo spirit.

A red dye is obtained from the wood.

*Arachis hypogaea* L.; Peanut, Groundnut, Monkey Nut

This twining herb has alternate, abruptly pinnate, compound leaves with bijugal leaflets, yellow flowers in axillary clusters, and subterranean pods. It is native to tropical America. Specimens were collected at Plymouth and Woodlands. *Brussell C-32, C-67.*

Peanuts are grown for food and were seen in the public market at Plymouth.

During the summer of 1977, the author noticed an aphrodisiac concoction made up of the chopped roots of David's root (*Chiococca alba* [L.] Hitchc.), chopped ramgoat roots (*Eryngium foetidum* L.), and whole peanut seeds soaked in a bottle of rum for sale at the Salem general grocery store.

*Caesalpinia bonduc* (L.) Roxb.; Grey Nicker, Horse Nicker; plate 14

This prickly rambling shrub or vine has alternate bipinnately compound leaves, racemes of yellow flowers, and brown, broadly elliptic-oblong, prickly pods containing one or two large shiny grey seeds. It is widely distributed in the New World and Old World tropics. Specimens were collected at Trants. *Brussell C-327.*

The roasted ground seeds are boiled and made into a drink that is taken to treat kidney trouble, heart ailments, and edema. The raw seeds are very poisonous and must be thoroughly parched before being used for medicinal purposes (Bayley 1949). The powdered raw seeds have been used for poisoning purposes on Montserrat.

Montserratians use the shiny grey seeds to make necklaces that have voodoo significance on the island. Necklaces incorporating the seeds are sold in the curio shops in Plymouth.

*Cajanus cajan* (L.) Huth; Pigeon Pea

This shrub has alternate pinnately three-foliate leaves, yellow flowers in axillary racemes, and brown compressed pods. It is native to the East Indies. Specimens were collected near Galway Soufriere and at Salem. *Brussell C-10, C-355.*

A tea made from the leaves is drunk to treat muscular strains and cold symptoms.

The seeds are eaten in the same manner as common garden peas and are high in protein.

*Canavalia ensiformis* (L.) DC.; Overlooker, Watchman

This climbing herb has alternate pinnately three-foliate leaves with large ovate leaflets. The large white and purple flowers produce large, long, wide, compressed pods. It is widespread in the tropics and subtropics of the world. The plant was found at Harris Village. *Sight record.*

The young pods are sliced and eaten like garden green beans.

The overlooker is planted around gardens because it is believed that it will make the garden crops bountiful and make other people jealous of the garden. This plant is also used as a voodoo border to keep thieves out of the garden.

*Cassia bicapsularis* **L.; Money Money Tree, Money Money Bush, Money Plant**

This shrub has alternate, abruptly pinnate, compound leaves with obovate-elliptic leaflets, yellow flowers in axillary racemes, and straight brown subcylindrical pods. It is native to the West Indies. Specimens were collected at Salem and in the White River Valley. *Brussell C-105, C-170, C-284.*

The leaves are given to pigs to treat gastrointestinal problems. The pulp in the pods is eaten raw or used to thicken stews.

*Cassia fistula* **L.; Golden Shower**

This tree has alternate, abruptly pinnate, compound leaves with ovate-oblong leaflets. The bright, yellow, showy flowers are borne in large pendulous clusters and give rise to large, brown, elongated, cylindrical pods containing numerous seeds and sweet gummy pulp. It is native to tropical Asia. Specimens were collected at Woodlands. *Brussell C-151.*

The sweet pulp of the fruit is eaten raw as a laxative.

*Cassia glandulosa* var. *swartzii* **(Wikstr.) J. F. Macbr.; Wild Tamarind**

This shrub has alternate, abruptly pinnate, compound leaves with glands. The small, yellow, solitary flowers give rise to brown pods with scattered hairs. It is native to the West Indies. The plant was found in the White River Valley. *Brussell C-277.*

The leaves are used to make bush tea for treating colds and nausea.

*Cassia obtusifolia* **L.; Money Bush**

This large herb has alternate, abruptly pinnate, compound leaves with obovate leaflets. The small yellow flowers produce elongated, compressed, curved pods. This Old World native was collected in the White River Valley. *Brussell C-53.*

A decoction made by boiling the entire plant is applied topically to the skin to treat "body pain."

*Cassia occidentalis* **L.; Stinking Bush, Wild Coffee**

This shrub has alternate, abruptly pinnate, compound leaves with ovate leaflets. The pale-yellow small flowers give rise to oblong-linear curved pods. It is native to the American tropics. Specimens were collected at Salem, Woodlands, and Olveston. *Brussell C-142, C-285, C-375*.

Tea made from the leaves is drunk to treat colds. An infusion made from the seeds is drunk as a treatment for jaundice (Duberry 1973). A paste made from the ground-up seeds and sweet oil is applied topically to cure ringworm. A decoction made from the fresh seeds is also used topically to treat ringworm.

The pulp of the fruit is eaten raw or cooked in sauces. The seeds are roasted and boiled to make a coffee substitute.

### *Cassia planisiliqua* L.; Wild Pea

This small tree has alternate, abruptly pinnate, compound leaves with elliptic leaflets, yellow flowers in axillary racemes, and elongated flat pods. It is native to the East Indies. A specimen was collected at Salem. *Brussell C-157*.

The roots are washed and chewed as a treatment for colds.

### *Centrosema virginianum* (L.) Benth.; Wild Blue Vine, Wild Pea, Bluebell

This slender twiner has alternate pinnately three-foliate leaves, blue to white flowers, and long narrow pods with black seeds. It is native to tropical and warm-temperate America. Specimens were collected at Woodlands and Cudjoehead. *Brussell C-121, C-335*.

After a thorough washing, the fresh roots are chewed and sucked on to treat sore throats. The fresh leaves are rubbed vigorously over the surface of the teeth to clean them after eating a meal. Tea made from the leaves and stems is said to "clean the blood."

### *Erythrina coralodendron* L.; Jumbie Bead Tree, Shrove Tuesday

This prickly tree has alternate pinnately three-foliate leaves, large showy red flowers borne in racemes, and brownish pods constricted between scarlet seeds with black markings. It is native to the West Indies. The tree was found in the White River Valley. *Sight record*.

The scarlet seeds of this tree contain a dangerous poison that is used to kill rats and insects. Necklaces and other decorative novelties are made from the seeds. "Jumbie beads" (the Montserratian name for the seeds) are used in voodoo practices on the island.

### *Galactia filiformis* Benth.; Strongback

This pubescent woody vine has alternate pinnately three-foliate leaves with ovate-oblong leaflets. The pink flowers give rise to linear pubescent pods. It is widely distributed in tropical America. Specimens were collected at Cudjoehead and Paradise. *Brussell C-186, C-338*.

Tea made by boiling the entire plant is drunk to treat back pain and "weakness of the bladder."

**Plate 1.** *Plumeria rubra*, red frangipani.

**Plate 2.** *Philodendron giganteum*, big chaney bush.

**Plate 3.** Coconut (*Cocos nucifera*) inflorescence.

**Plate 4.** Informant Patricia Rabess with coconut (*Cocos nucifera*) leaf fibers to be used for making a broom.

**Plate 5.** *Crescentia cujete*, calabash.

**Plate 6.** *Bixa orellana*, annatto.

**Plate 7.** Dugout canoe, near completion, made from a silkcotton tree (*Ceiba pentandra*).

**Plate 8.** Informants Daniel Allen and James Cabey with fruits of *Mammea americana*, mammy apples.

**Plate 9.** The author collecting specimens of *Ipomoea pes-caprae* subsp. *brasiliensis*, beach morning glory.

**Plate 10.** *Momordica charantia*, maiden apple. Derivatives from this plant have been shown to inhibit replication of the AIDS virus, herpes simplex virus I, and poliovirus I, as well as to exhibit anti-tumor and immune enhancement ability and to inhibit prostate adenocarcinoma in rats and lymphoma in mice (Lee-Huang et al. 1990).

**Plate 11.** *Sechium edule*, christophine.

**Plate 12.** Informant holding a specimen of *Diospyros revoluta*, black apple. The seeds and bark chips are used to stupefy fish.

**Plate 13.** *Hura crepitans*, sandbox tree. The seed capsules may explode with a startling noise when mature.

**Plate 14.** *Caesalpinia bonduc*, grey nicker. The highly toxic seeds are used as poisoning agents.

**Plate 15.** *Piscidia piscipula*, dogwood.

**Plate 16.** *Hibiscus rosa-sinensis*, hibiscus.

**Plate 17.** *Hibiscus sabdariffa*, sorrel.

**Plate 18.** *Artocarpus altilis*, breadfruit.

**Plate 19.** Informant Fred Payne with *Castilla elastica*, wild rubber.

**Plate 20.** *Heliconia caribaea*, balisier.

**Plate 21.** Informant James Lee cutting *Saccharum officinarum*, sugarcane.

**Plate 22.** *Coccoloba uvifera*, seagrape.

**Plate 23.** *Mimusops coriacea*, Aphrodite's apple, is eaten as an aphrodisiac.

**Plate 24.** *Datura suaveolens*, angel's trumpet.

***Gliricidia sepium*** (Jacq.) Steud.; Rainfall Tree, Gorey Cedar

This tree has alternate pinnately compound leaves with an odd leaflet, small pink flowers, and elongated flattened pods. It is native to Central America. The tree was found at Salem. *Brussell C-177, C-299.*

Tea made from the leaves is drunk to treat colds, coughs, and fevers. The freshly crushed leaves are used in poultices.

The fresh blossoms are stir-fried and eaten. The toxic seeds and roots are used to poison rats, mice, dogs, and other animals. The hard trunks are set out as living fence posts.

***Haematoxylon campechianum*** L.; Logwood, Dyewood

This tree has alternate, abruptly pinnate, compound leaves with obovate leaflets, small yellow flowers in axillary racemes, and flat oblong winglike pods. It is native to Yucatan. Specimens were collected at Olveston. *Brussell C-248.*

Tea made by boiling chips of the sweet-smelling heartwood of this tree is drunk to treat dysentery.

Dark blue dye is extracted by boiling chips of the heartwood and used to color clothing dark navy blue. This dye is also used to color steel traps in order to camouflage them and to help prevent rusting.

***Hymenaea courbaril*** L.; Locust Tree, Sweet Pod

This tree has alternate leaves consisting of a pair of ovate oblique leaflets, white flowers, and thick brown subcylindrical pods. It is native to the West Indies. Specimens were collected at Salem and in the White River Valley. *Brussell C-103, C-295.*

The seeds are parched, ground, and boiled in water to make a drink that is a treatment for jaundice. A decoction made by boiling the bark is drunk to treat chest colds.

The sweet mealy pulp found in the pods is eaten raw or cooked in sauces and stews. After having eaten the raw pulp contained in one large pod, I found it to have a stimulating and slightly hallucinogenic effect.

A beverage is made by steeping the pulp in hot water.

The gum that exudes from the trunk and roots is burned as incense in voodoo ceremonies and in some religious ceremonies. This gum is also used to caulk boats.

***Indigofera suffruticosa*** Mill.; Indigo Weed

This shrub has alternate pinnately compound leaves with oblong leaflets, purple flowers, and narrow subcylindrical sickle-shaped pods. It is native to the West Indies. The plant was found in the Galway Soufriere area. *Brussell C-29.*

Tea made by steeping the leaves is drunk as a purgative. Poultices made from the crushed fresh leaves are applied to wasp stings.

*Inga laurina* (Sw.) Willd.; Spanish Oak

This tree has alternate, abruptly pinnate, compound leaves with oblong-linear leaflets, many white funnelform flowers, and oblong flattened pods. It is native to the West Indies. Specimens were collected at Woodlands. *Brussell C-132, C-282.*

The leaves are steeped to make a beverage tea. Children suck the sweet pulp from the large seeds of this tree. The tough, sturdy, easily worked wood is used for furniture, cabinets, interior trim, and charcoal.

*Mimosa pudica* L.; Shame Bush, Sleepy, Strongback

This rambling shrub has recurved prickles, alternate bipinnately compound leaves that fold when touched, pink flowers in globose heads, and bristly brown linear-oblong pods. It is native to tropical America. Specimens were collected in the White River Valley. *Brussell C-58, C-102.*

Tea made by boiling the leaves is drunk to treat back pain and menstrual cramps. A decoction made by boiling the entire plant is used to treat colds.

*Mucuna pruriens* (L.) DC.; Cow Itch Vine

This twining vine has alternate pinnately three-foliate leaves, large purple flowers, and dark brown pods covered with stinging hairs. It is widespread in the tropics. A large population of these plants was found in the White River Valley. *Sight record.*

Tea made by boiling the leaves is drunk as a cold medicine.

*Phaseolus lunatus* L.; Lima Bean, White Bean

This perennial twiner has alternate pinnately three-foliate stipulate leaves and white racemose flowers that produce compressed, elongated, lunate pods that contain flattened kidney-shaped seeds. It is native to tropical America. Lima beans were found in the public market at Plymouth. *Sight record.*

The cooked beans are commonly eaten on Montserrat and are a good source of protein.

*Phaseolus vulgaris* L.; Kidney Bean

This twining herb has alternate, pinnately three-foliate, stipulate leaves and racemose flowers that produce elongated cylindrical pods that contain reddish-brown kidney-shaped seeds. It is native to tropical America. Kidney beans were found in the public market at Plymouth. *Sight record.*

Kidney beans serve as a proteinaceous food source on the island.

*Piscidia carthagenensis* **Jacq.; Fish Poison Tree**

This tree has alternate pinnately compound leaves with oblong leaflets that narrow into a minute point, numerous purplish to pinkish-white flowers in panicles, and brown indehiscent pods with four longitudinal wings. It is native to the West Indies. Specimens were collected at Harris Village. *Brussell C-332.*

Tea made from the root bark is drunk to induce sleep, ease cold symptoms, and relieve pain.

The roots and branches are pounded with stones and vigorously shaken in a stream. Piscidin, with its hypnotic and sedative properties, is leached into the water causing fish to be stupefied and float to the surface. Fishermen gather the incapacitated fish by hand or catch them in nets that are strung across the body of water downstream. Alfred Payne of Salem said that when the fish poison tree is used in a marine inlet or bay, fishermen throw chopped or pounded young branches, root bark, and fresh or dried leaves into the water. Another method related by Mr. Payne involves putting the ground-up root bark, branches, and leaves into burlap bags that are dragged in circles by swimmers or boats. The stupefying effect on the fish is temporary. Those that are not caught recover.

The tough, hard, heavy wood is used for making vehicles, boats, bridges, and fence posts.

*Piscidia piscipula* **(L.) Sarg.; Dogwood; plate 15**

This small tree has alternate pinnately compound leaves with elliptic leaflets, numerous white flowers in panicles, and brown indehiscent pods with four longitudinal wings. It is native to the West Indies. Specimens were collected at Harris Village and Great Alp Falls. *Brussell C-55, C-202, C-331.*

Piscidin, which has hypnotic and sedative properties, is found in the entire tree, but is most highly concentrated in the root bark. Ground-up pieces of the root bark are put in the cavity of a decayed tooth to relieve the pain of toothache (Gooding et al. 1965). On Montserrat tea made by boiling pieces of the root bark is imbibed to induce sleep and relieve pain with no apparent side effects. This same tea is also taken to treat cold symptoms.

The crushed root bark is used as a fish intoxicant and as an insecticide. A decoction of the macerated root bark is used as an external wash to cure mange on dogs as well as to rid them of fleas and ticks. The hard strong wood is used for tool handles and fence posts.

*Pithecellobium arboreum* **(L.) Urban; Wild Tamarind**

This tree has alternate bipinnately compound leaves, numerous white flowers in globose heads, and conspicuous red coiled pods. It is native to the West Indies. The tree was found in the White River Valley. *Sight record.*

The leaves are boiled, and the resulting decoction is drunk to treat nausea and colds.

The strong durable wood is used for cabinets, furniture, interior work, and general construction.

### *Pithecellobium dulce* (Roxb.) Benth.; Sweet Tamarind

This tree has alternate bipinnately compound leaves, small white flowers, and curved brown pods with shiny black seeds. It is native to tropical America. The tree was found at Olveston. *Sight record.*

The sweet piquant pulp surrounding the seeds is eaten raw or dissolved in water to make a beverage similar to lemonade.

### *Pithecellobium saman* (Jacq.) Benth.; Raintree, Rainfall Tree

This large tree has alternate bipinnately compound leaves with numerous diamond-shaped leaflets, pink brushlike flower clusters, and flattened elongated pods. It is native to tropical America. Specimens were collected at Olveston and White River Ghaut. *Brussell C-50, C-96, C-350.*

Tea made by steeping the dried leaves is drunk to treat fevers, coughs, colds, and asthma.

The pods are eaten raw and have a flavor similar to licorice. Cattle, hogs, and goats are fond of the fruits and serve as vectors for distributing the seeds. The trunks are often used for fence posts.

### *Pithecellobium unguis-cati* (L.) Mart.; Bread and Cheese

This small tree has alternate bipinnately compound leaves, pink flowers in globose heads, and flattened twisted pods with fleshy red arils around the black seeds. It is native to tropical America. Bread and cheese trees were found on the McChesney Estate at Olveston. *Brussell C-251.*

The fleshy arils are eaten raw or cooked in soups. Necklaces are made from the seeds.

### *Prosopis juliflora* (Sw.) DC.; Cashee

This tree has stipular spines, alternate bipinnately compound leaves, yellow flowers in cylindrical spikes, and compressed falcate pods. It is native to continental tropical America. *P. juliflora* was found at Woodlands. *Sight record.*

The sap and stem exudate is applied topically to cuts to reduce scar-tissue formation.

The wood is used for charcoal.

### *Pterocarpus indicus* Willd.; Burma Rosewood

This giant tree has alternate pinnately compound leaves with ovate leaflets, fragrant yellow flowers, and nearly round brown pods bordered with a thin wing. It is native to southeast Asia. Specimens were collected at Shoe Rock. *Brussell C-255.*

An infusion made from the bark and chips of the wood is drunk as a diuretic.

The wood is used for general construction and is highly valued for furniture. Cups carved from the wood give water a wonderful blue color that changes due to variations in light intensity.

*Tamarindus indica* L.; Tamarind, Tamarind Bush, Taman

This tree has alternate abruptly pinnate compound leaves, yellow flowers, and brown curved pods with acidic pulp surrounding the seeds. It is native to the Old World and has been widely naturalized in the tropics. Specimens were collected at Cudjoehead and White River Ghaut. *Brussell C-45, C-341.*

Tea brewed from the fruits and leaves is drunk to treat coughs, colds, and fevers. Hot water is poured over the leaves. The resulting infusion is allowed to steep for five minutes and then drunk to relieve lung congestion and asthma. The fresh leaves and immature fruits are chewed and sucked on to treat a sore throat. The mature fruits are shelled and the brown pulp is eaten to stimulate the appetite and for their laxative effect.

"Taman soup" is made by stewing the fruits in syrup. The tart pulp of the fruits is used for seasoning purposes.

## FLACOURTIACEAE

*Flacourtia jangomas* (Lour.) Raeusch.; Governor's Plum, Java Plum

This tree has alternate ovate leaves, small greenish-white flowers in racemes, and juicy drupes. It is native to the Old World tropics. Specimens were collected at Paradise and Olveston. *Brussell C-194, C-254.*

The fruits are "rimed" (rubbed between the thumb and forefinger) and eaten raw. These fruits are also used in preserves, sauces, and pies.

## LAMIACEAE

*Leonotis nepetifolia* (L.) Ait.; Lord Lovington

This stout herbaceous plant has opposite ovate leaves, yellow-orange flowers, and smooth nutlets. It is native to the Old World and has been widely naturalized in the tropics and subtropics. Specimens were collected at Woodlands. *Brussell C-137, C-145.*

The leaves are steamed and applied topically with cloth wrapping to draw out boils.

*Mentha piperita* L.; Peppermint

This aromatic purple-stemmed herb has opposite ovate leaves, terminal clusters of purple flowers, and small dry spherical fruits. It is native to the Old World. Specimens were collected at the public market at Plymouth. *Brussell C-69.*

Tea made from the leaves is drunk to relieve cold symptoms and to treat an upset stomach. This stimulating tea is also drunk as a refreshing beverage.

The leaves are used for seasoning purposes.

### *Mentha viridis* L.; Spearmint

This quadrangular-stemmed aromatic herb has opposite elliptic-oblong leaves, purple flowers in axillary whorls, and small ovoid dry fruits. It is native to the Old World. Spearmint was for sale at the Plymouth public market. *Sight record.*

Tea made from the leaves is drunk to relieve indigestion and gastrointestinal distress, including flatulence. The tea is also drunk as a cooling agent for reducing fevers.

The leaves are used fresh in salads and dried as a flavoring agent in cookery.

### *Nepeta cataria* L.; Catmint, Catnip

This pubescent herb has opposite cordate-ovate leaves, white flowers, and small dry ovoid fruits. It is native to the Old World. Specimens were collected at the Galway Soufriere, Plymouth, and St. George's Hill. *Brussell C-30, C-65, C-307.*

A strong tea made from the leaves and stems is drunk to "throw away picknies" (induce abortions), while a less potent brew is drunk to expel intestinal worms and to treat colds and stomach distress.

The young shoots and leaves are used for seasoning purposes.

### *Ocimum micranthum* Willd.; Duppy Basil, Mosquito Bush

This branched herb has opposite ovate leaves, small lavender flowers, and shiny black nutlets. It is native to tropical America. Duppy basil was found at Salem. *Sight record.*

The entire plant is crushed and hung in homes to repel mosquitoes. "Duppy" is a Montserratian obeah term used to denote a ghost or a mosquito (cf. Dobbin 1986).

### *Origanum majorana* L. ( = *Majorana hortensis* Moench); Marjoram, Sweet Marjoram

This suffrutescent herb has opposite ovate leaves, pink flowers borne in rounded heads, and small globose dry fruits. It is native to the Mediterranean area. Specimens were collected at the public market at Plymouth. *Brussell C-68.*

Tea made from the leaves and stems is drunk to treat colds, lung congestion, colic, and nausea. A decoction made from the leaves, stems, and flowers is applied topically to treat "strains" (of muscles). According to Morton (1981), a decoction of the leaves and stems is taken as an expectorant in Venezuela.

The dried leaves are used for seasoning purposes.

*Origanum vulgare* L.; Sweet Marjoram, Oregano, Balsam

This small suffrutescent herb has opposite ovate leaves, purple flowers set in oblong heads, and dry globose fruits. It is native to the Old World. Specimens were obtained at Salem and at the public market in Plymouth. *Brussell C-60, C-64, C-178.*

Tea made from the leaves is drunk to treat colds and to promote menstrual flow.

The leaves are used fresh or dried to flavor food.

*Rosmarinus officinalis* L.; Rosemary

This suffrutescent herb has opposite linear obtuse leaves, light blue flowers, and dry ovoid fruits. It is native to the Mediterranean area. Specimens were obtained at the public market in Plymouth. *Brussell C-62.*

Tea made from the leaves is drunk to cure headache. A decoction made from the entire plant is mixed with borax and used as a hair rinse to prevent baldness.

*Thymus vulgaris* L.; Thyme, Ticky Thyme; figure 35

This small aromatic shrub has opposite oblong-ovate leaves, purple to white flowers, and tiny globose dry fruits. It is native to the Mediterranean area. Specimens were collected at Paradise. *Brussell C-262.*

A tea made from the leaves and stems is drunk to treat coughs, colds, headaches, and hangovers. The most effective cough syrup I ever encountered con-

Figure 35. Informant Daniel Allen (*left*) holding a thyme plant (*Thymus vulgaris*) at the edge of a companion planting of banana (*Musa sapientum*) and thyme. The thyme is planted to repel insects from the banana plants and as a second cash crop.

tained thyme extract as its active ingredient. A decoction made by boiling the entire plant is drunk as a diuretic, to induce menstrual flow, and in stronger doses to induce abortion.

The dried leaves are used to flavor foods.

On Montserrat thyme is widely grown both for local use and for export. I observed large thyme gardens growing between rows of bananas at Paradise.

## LAURACEAE

### *Cassytha filiformis* L.; Love Vine, Love, Yellow Dodder

This climbing parasitic herb has alternate leaves reduced to small scales, small white flowers, and globose white fleshy fruits. It is widespread in the tropics. Specimens were collected on the McChesney Estate at Olveston. *Brussell C-247*.

Tea made from the entire plant is drunk to treat colds and as a purificant.

A decoction made by boiling the leaves and stems together with the bark of *Bursera simaruba* (L.) Sarg. is drunk as an aphrodisiac that is believed to help win the love of a member of the opposite sex.

The one who imbibes this potion is said to become amorously stimulated and is also believed to become more appealing to the person the drinker wishes to influence.

### *Cinnamomum zeylanicum* Bl.; Cinnamon

This aromatic tree has opposite ovate leaves, and clusters of small white flowers that produce elliptic black berries. It is native to Ceylon. Specimens were collected at the McChesney Estate at Olveston. *Brussell C-245*.

Tea made from the leaves, twigs, and bark is drunk as a treatment for indigestion and diarrhea.

The bark is used as a flavoring agent.

### *Licaria triandra* (Sw.) Kosterm.; Sweetwood

This tree has alternate ovate leaves, red flowers, and dark blue drupes with thin flesh. It is native to tropical America. Specimens were collected at Woodlands. *Brussell C-115*.

Tea made from the bark is drunk to treat gastrointestinal distress.

The wood is used for interior construction and to make charcoal.

### *Nectandra coriacea* (Sw.) Griseb.; Sweetwood, Black Torch, Lancewood

This tree has alternate narrowly elliptic leaves, small white flowers in panicles, and black globose berries. It is native to the American tropics. A sweetwood tree was found at Woodlands. *Brussell C-172*.

The leaves are aromatic when crushed and are steeped to make a tea that is drunk as a treatment for colds and nausea.

The wood is used for cabinets, general carpentry, and for poles.

*Nectandra membranacea* (Sw.) **Griseb.;** Southernwood

This tree has alternate elliptic to lanceolate leaves, small yellowish-white flowers in panicles, and small black globose berries. It is native to the West Indies. Specimens were purchased at the Plymouth public market. *Brussell C-61.*

The sweetly aromatic leaves are used to make an appealing bush tea beverage.

*Persea americana* **Mill.; Avocado**

This tree has alternate elliptic leaves, greenish-yellow hairy flowers, and heavy green pear-shaped, round, or oval fruits with thick buttery flesh. It is native to tropical America. Avocado trees were found growing at Woodlands. *Sight record.*

The mashed pulp of the ripe fruits is used as cream for softening dry skin, in poultices, and as a facial mask. The ripe fruits are eaten raw fresh off the tree, cut up in salads, and mashed in dips. Avocados are sold in the public market at Plymouth.

## LILIACEAE

*Allium fistulosum* **L.; Chivel, Welsh Onion**

This erect herb has linear, long, narrow, inflated leaves and small, white, cup-shaped flowers that give rise to pinkish-green three-lobed capsules. It is native to Asia. Specimens were collected at St. George's Hill. *Brussell C-190.*

A tea made by boiling the leaves is drunk as a vermifuge, diuretic, and diaphoretic.

The bulbs and leaves are eaten raw and cooked.

*Allium sativum* **L.; Garlic**

This herb has flat, long, linear leaves and white cup-shaped flowers that produce green, lobed, three-valved capsules. It is native to the Old World. Garlic bulbs were found in the public market at Plymouth. *Sight record.*

The bulbs are simmered, and the resulting decoction is drunk to dispel stomach gas, expel intestinal worms, and treat bladder trouble, coughs, asthma, and hoarseness. Crushed garlic bulbs are rubbed on wasp stings.

The bulbs are used for seasoning purposes in cookery.

*Aloe vera* (L.) **Burm. f.; Sintibibi, Aloe, Bitter Aloe; figure 36**

This succulent herb has fleshy swordlike leaves with sharply toothed margins, yellow tubular flowers, and leathery capsules with many black seeds. It is native to the Mediterranean area. Specimens were collected in the White River Valley. *Brussell C-279.*

**Figure 36.** *Aloe vera*, sintibibi, is used to treat stomach ulcers.

Aloe juice squeezed from the leaves is mixed with salt and lime juice, heated, cooled, and applied to boils, cuts, and burns. Aloe juice mixed with salt and lime juice is also used as a gargle for a sore throat and is taken in two-tablespoon doses four times a day to treat diabetes. Aloe juice is taken internally when the "moon is weak" as a cure for jaundice and is also used as a shampoo.

Aloe leaves are split, heated over a fire, and rubbed on a sprain with some salt to help speed up the recovery process and to ease the discomfort. The split, heated leaves are also applied topically to sea urchin punctures. After leaving the leaf poultice on the wound for one day the sea urchin spine can be easily removed. The aloe leaf poultice also reduces pain from sea urchin wounds. Aloe poultices are applied to wasp stings to reduce swelling.

Pulp from the leaves is mixed with salt and water and taken internally to prevent and to treat cancer. The peeled and salted leaves are chewed and swallowed as a laxative. The macerated pulp and juice from the leaves is mixed with wine and given to children as a vermifuge beverage in one cup doses morning and evening for one week.

Aloe gel squeezed from the leaves is taken internally for stomach trouble and is said to cure a bleeding ulcer in five to ten days when taken in doses of two tablespoons four times a day (one hour after each meal and before bedtime).

Aloe leaf gel is very effective for treating sunburn, as I know from experience. This gel is regarded by Montserratians as an agent that speeds healing and prevents infection.

*Sansevieria metallica* Gerome & Labroy; Coretor, Skip Rope, Skip Vine, Bowstring Hemp

This perennial herb has lanceolate sword-shaped leaves with a fine red line bordering the margin, greenish white flowers, and thin-walled capsules with fleshy seeds. It is native to the Old World tropics. Specimens were collected at Lime Kiln Beach. *Brussell C-155.*

The juice from the leaves is applied to wounds to reduce scar tissue formation and also is applied to old scars to diminish them.

Children pound the leaves with stones and make skip ropes by tying several of them together. I observed this playful pastime during the Christmas season.

## LOGANIACEAE

*Spigelia anthelmia* L.; Worm Grass

This herb has opposite leaves, small purple flowers, and two-lobed capsules with many seeds. It is native to tropical America. Worm grass was found growing near the Emerald Isle Hotel. *Sight record.*

A decoction made by boiling the entire plant is drunk as an anthelmintic.

## LORANTHACEAE

*Phoradendron trinervium* (Lam.) Griseb.; No Mammy, Mistletoe; see figure 23

This parasitic shrub has opposite obovate leaves, small yellow-green flowers borne in spikes, and transparent, yellow-orange, ovoid, glutinous berries. It is native to the West Indies. Specimens were collected in the Centre Hills. *Brussell C-91.*

Tea made from the leaves and twigs is drunk to expel a "pickney" (fetus) or to prevent conception. The daily dose for birth control is one cup of the tea in the morning followed by another in the evening. A larger dose is taken to induce abortion. The author does not recommend this practice due to the poisonous properties associated with the genus *Phoradendron* (Brussell 1977). An overdose could cause death.

The plant is also used in voodoo.

## MALPIGHIACEAE

*Byrsonima spicata* (Cav.) DC.; Shoemaker Bark

This forest tree has opposite elliptic leaves, showy yellow flowers, and globose yellow fruits. It is native to tropical America. The tree was found growing in the Centre Hills. *Brussell C-163.*

A decoction made by boiling the bark is used to tan leather.

*Malpighia punicifolia* L.; Barbados Cherry, Acerola Cherry

This small tree has opposite elliptic leaves, small pink flowers, and scarlet globose drupes. It is native to the West Indies. Specimens were collected on the Tar River Estate and at Olveston. *Brussell C-72, C-244.*

The fruit, which is very high in vitamin C content, is eaten raw or cooked in preserves.

## MALVACEAE

*Abutilon hirtum* (Lam.) Sweet; Burry Bark

This softly tomentose shrub has alternate, partly lobed, cordate leaves and yellow and red flowers that give rise to brown fruits containing three seeds each. It is widespread in tropical areas of the world. Specimens were collected at Paradise. *Brussell C-191.*

The tough fibrous bark is used to make rope.

*Abutilon indicum* (L.) Sweet; Rope Bush

This tomentose shrubby herb has alternate rounded-cordate leaves and yellow flowers that produce dry brown fruits with three seeds each. It is widespread in tropical areas of the world. Specimens were collected near the Galway Soufriere. *Brussell C-9.*

The stem fibers are used to make strong rope.

*Gossypium barbadense* L.; Sea Island Cotton; figure 37

This shrub has alternate three- to five-lobed leaves, solitary yellow flowers, and coriaceous capsules containing seeds covered with long white hairs. It is native to tropical America. Specimens were collected in a cultivated field in the White River Valley. *Brussell C-274.*

The root bark is boiled, and the resulting decoction is drunk to treat dysmenorrhea and to suppress the menses. This decoction is said to cause abortion when taken in large doses.

Sea island cotton was once an important crop on the island. It is still grown sporadically but is no longer a major crop.

*Hibiscus rosa-sinensis* L.; Hibiscus; plate 16

This shrub has alternate ovate leaves, large red, purple, white, or yellow flowers, and capsules containing many seeds. It is native to tropical Asia. Specimens were collected at Woodlands. *Brussell C-301.*

Tea made from the dried flowers is drunk to treat colds. A decoction made by boiling the leaves is mixed with lemon juice and drunk to treat a sore throat.

The flowers are rubbed and macerated in water until the mixture becomes thick. The fluid is then strained and used as a hair wash.

**Figure 37.** Informant demonstrating ridge tillage method for planting sea island cotton (*Gossypium barbadense*).

*Hibiscus sabdariffa* L.; Sorrel, Roselle; plate 17

This suffrutescent herb has alternate palmate-veined leaves with stellate hairs, red flowers, and five-locular capsules. It is widespread in the tropics. Specimens were collected at Salem and Woodlands. *Brussell C-123, C-231, C-354.*

The fleshy red calyces are boiled fresh in water with sweetening and spices to make a piquant burgundy-colored Christmas libation. I found this drink to be an enjoyable, colorful, yuletide beverage when sweetened with Barbados honey. It can be enjoyed either hot or cold. The calyces are stewed and made into a jam that is somewhat reminiscent of cranberry sauce (Herman 1973). Wine is also made from the calyces. Some Montserratians prefer to place crushed calyces and chopped ginger rhizomes in a container along with water and molasses and allow the mixture to ferment, thus producing a piquant alcoholic beverage.

Sorrel calyces are sold in the public market in Plymouth at Christmas.

*Hibiscus vitifolius* L.; Wild Okra, Rope Plant

This shrub has alternate three- to five-lobed leaves, yellow flowers with a purple base, and globose five-winged capsules. It is native to the Old World tropics. Specimens were collected in the White River Valley. *Brussell C-276.*

The tough fibrous bark is used to make rope.

*Sida ciliaris* L.; Twelve O'clock

This suffrutescent herb has alternate oblong leaves, orange flowers, and schizocarps with one seed per carpel. It is native to tropical America. Specimens were collected above Salem and in the White River Ghaut. *Brussell C-44, C-128, C-166.*

The leaves and soft stems are pounded with stones, mixed with salt, and tied on sprained ankles and other swollen areas as a poultice to reduce pain and edema.

*Thespesia populnea* (L.) Soland; Seaside Mahoe

This spreading tree has alternate cordate leaves, large yellow and violet flowers, and leathery brown fruits. It is widely distributed on tropical shores around the world. Seaside mahoe was found growing at Lime Kiln Beach. *Sight record.*

A decoction made by boiling the fruits is applied externally to treat skin eruptions.

The flowers are fried and eaten. Rope is made from the tough fibrous bark.

## MARANTACEAE

*Maranta arundinacea* L.; Arrowroot

This erect herb has alternate ovate leaves and white tubular flowers that give rise to broad-oblong three-sided capsules that contain pink seeds, each flanked by a yellow aril. It is native to South America. Arrowroot was found growing at Woodlands. *Sight record.*

The young rhizomes are roasted or boiled and eaten as vegetables. Starch is extracted from the tuberous carrot-shaped rhizomes of this plant. The arrowroot starch produced on Montserrat is mostly used for local consumption at present. However in the early part of this century, arrowroot starch was "produced in large quantities and [was] of very superior quality" for export (Shafer 1907).

## MELASTOMACEAE

*Miconia prasina* (Sw.) DC.; Hogwood, Camasey

This small tree has opposite ovate-lanceolate leaves, small white flowers in panicles, and slightly flattened reddish-blue berries. It is native to the American tropics. Specimens were collected at the base of Chance Peak and near the Galway Soufriere. *Brussell C-271, C-365.*

The juicy slightly tart berries are eaten raw or cooked in pies and preserves.

## MELIACEAE

*Cedrela mexicana* M. J. Roem.; Red Cedar; figure 38

This tree has alternate abruptly pinnate leaves with ovate-lanceolate leaflets, small green flowers in

## Ethnobotanical Uses and Specific Discussion

**Figure 38.** Small boat and fish traps made of red cedar (*Cedrela mexicana*) at Carr's Bay.

terminal panicles, and five-celled capsules. It is native to the West Indies. Specimens were collected near Great Alp Falls and at Olveston. *Brussell C-59, C-93.*

A decoction made by boiling the leaves is used as a perspiration-inducing bath. The leaves and branches are wrapped around a person with a fever to draw the heat out. According to Duberry (1973) the leaves are spread under the bedsheet of a fever patient. Lying on these leaves is said to make the fever go away. An infusion of pulverized leaves mixed with salt and warm water is drunk to treat fish poisoning.

The reddish-brown aromatic wood is nearly impervious to insects and is highly valued for furniture, cabinets, woodwork, fish traps, and boatbuilding. During my first stay on the island, a 16 m wooden cargo boat was built and launched at Fox's Bay. Most of the wood used to build this boat was red cedar.

### *Melia azadirachta* L.; Neem

This tree has alternate pinnate leaves and numerous, small, white flowers that give rise to yellow elliptic drupes. It is native to southern India. A neem tree was found growing in Plymouth. *Sight record.*

Tea made by steeping the dried leaves is taken by women during pregnancy (Duberry 1973).

A decoction made from the leaves, fruits, and bark is sprayed on garden plants to combat insects. The wood is used for fence posts and general construction.

*Melia azedarach* L.; Lilac, Redwood

This tree has alternate bipinnate leaves with ovate-lanceolate leaflets, fragrant lavender-colored flowers, and yellow ellipsoid drupes. It is native to tropical Asia. Specimens were collected at Woodlands and in White River Ghaut. *Brussell C-47, C-112, C-116.*

Montserratians boil the leaves and apply the resulting decoction topically to relieve pain. The tea made by boiling the leaves is taken as a cold treatment. The leaves are placed in a pillow, and the patient then sleeps with his or her head on the pillow to treat a fever.

The fruits are poisonous, and hogs have died from eating them. The wood is favored for making charcoal.

*Swietenia macrophylla* King; Honduras Mahogany

This large deciduous tree has alternate, even, pinnate leaves with unequal-sided elliptic to oblong leaflets and fragrant greenish-yellow flowers that give rise to erect, egg-shaped, woody capsules that split into five parts. It is native to mainland tropical America. Specimens were collected at Woodlands. *Brussell C-200.*

The flowers of *S. macrophylla* are an important source of honey, and a decoction made by boiling its bark is used topically as an astringent.

Honduras mahogany is highly esteemed for cabinets, furniture, woodwork, and tool handles. This is one of the most valuable lumber trees in the American tropics. A number of Honduras mahogany logs were observed stacked and ready to be sawed at the Montserrat sawmill.

*Swietenia mahagoni* (L.) Jacq.; Mahogany; figure 39

This deciduous tree has alternate abruptly pinnate leaves with lanceolate leaflets, greenish-yellow flowers, and woody five-valved capsules. It is native to the American tropics. Specimens were collected at Bugby Hole. *Brussell C-328.*

The leaves are used to make a decoction that is drunk to treat colds and fevers.

Mahogany is a valuable timber tree on Montserrat that is used for general construction, furniture, cabinets, coffins, and tool handles. The reddish-brown heartwood is resistant to decay and attack by termites, but the sapwood is said to be very susceptible to insects and decay.

While on a collecting expedition in the Centre Hills, Daniel Allen and I came upon an upper level banana garden in the rainforest. Overhead, a well-made mahogany coffin was hanging from a tall tree. The coffin, which was suspended 10 m above the ground, served as an eerie voodoo symbol warning people not to take bananas from the garden.

## MORACEAE

*Artocarpus altilis* (Parkinson) Fosb. ( = *Artocarpus communis* J. R. & G. Forst.); Breadfruit; plate 18

This tree has alternate, deeply cut, elliptic leaves and yellowish-green flowers that give rise to large, globose, green to brown, multiple fruits. It is native to the South Pacific Islands. Specimens were collected at Woodlands. *Brussell C-130.*

Tea made by boiling yellow leaves picked from the trees is drunk to lower blood pressure and as a cure for heart trouble.

The green fruits are baked and eaten like bread or boiled and used like common potatoes. In fact, the taste of green breadfruit is somewhat like that of the Irish potato with an added nutty flavor. Breadfruit au gratin casserole has an exquisite taste and became one of my favorite dishes on the island. The ripe fruits become brown, soft, and sweet and are cooked and used to make puddings. Breadfruit blossoms are cooked and candied (Herman 1973).

The white latex from this tree is used as glue.

The famous Captain Bligh brought this valuable tropical tree to the St. Vincent Botanical Garden in the West Indies from Tahiti in 1793. Breadfruit trees were then distributed throughout the West Indies as a cheap starchy food for slaves that would take the place of bread made from costly imported grain.

### *Artocarpus heterophyllus* Lam.; Jackfruit

This tree has alternate oblong leaves, yellowish-green flowers, and giant elliptic yellowish-green com-

Figure 39. Maghogany coffin suspended from a tall tree serving as a voodoo symbol warning people not to take bananas from the garden below.

pound fruits. It is native to tropical Asia. Specimens were collected at Olveston on the McChesney Estate. *Brussell C-249.*

The unripe fruits are sliced and cooked like potatoes. The sweet fleshy "bulbs" of the ripe fruits are eaten raw or preserved in syrup.

***Castilla elastica*** **Cervantes; Wild Rubber Tree; plate 19**

This tree has pubescent, alternate, oblong leaves and clusters of greenish-yellow flowers that give rise to multiple disk fruits. It is native to continental tropical America. Specimens were collected on Chance Peak, in the Centre Hills, and on Katy Hill. *Brussell C-81, C-223, C-324.*

This tree was once an important source of rubber on Montserrat. Occasionally these trees are still tapped to get rubber for local uses such as making balls or waterproofing shoes, boats, and roofs.

***Cecropia peltata*** **L.; Trumpeter Tree, Snakewood; figure 40**

This tree has alternate, palmately lobed, peltate leaves and small yellowish-green flowers that give rise to cylindrical, fleshy, multiple fruits. It is native to the West Indies. Specimens were collected at Salem. *Brussell C-232.*

The leaves are dried, finely crushed, and steeped in boiling water. The resulting tea is drunk to treat high

**Figure 40.** *Cecropia peltata*, trumpeter tree.

blood pressure, colds, asthma, bronchial congestion, and congestion of the sinuses (Hodge and Taylor 1957).

A decoction made by boiling the bark and latex of this tree is mixed with honey or molasses and used as cough syrup.

Fishnet floats and bottle stoppers are made from the soft light wood. Flutes are made from the hollow petioles.

### *Ficus aurea* Nutt.; Evil Tree

This spreading tree has alternate ovate leaves, tiny inconspicuous flowers borne on the inner walls of globose, hollow, fleshy receptacles, which develop into small, spherical, fleshy, red fruits. It is native to the West Indies. Evil trees were found at Woodlands. *Brussell C-225*.

The fruits are sometimes eaten raw but have very little taste. The milky sap is a source of rubber and is used to caulk boats.

Montserratians call this plant "the evil tree" due to its habit of spreading and literally strangling other trees in its vicinity. People of the island believe that evil spirits called "Jumbies" live in the roots of this tree and that if a person cuts or harms the tree, these Jumbies will "get after" that person. It is for this reason that these trees are seldom cut by Montserratians.

A Caucasian expatriot living on the island tried to get his hired hand to cut an evil tree, but the worker flatly refused. Several other Montserratians were asked to cut the tree but also refused. The expatriot then decided to cut it himself but after some contemplation felt he did not want to cut it either. As of my last visit, the evil tree was still standing.

An elderly man from Woodlands told me Jumbies were once common all over the island and could sometimes be seen in broad daylight. However, after the appearance of cars, paved roads, and more people and houses, the Jumbies moved up into the heavily forested hills or into the dense aerial roots of the evil trees. The Jumbies are said to only come down to the settled areas at night when everything is quiet and are only occasionally seen at daybreak. This fanciful tale is reminiscent of stories told about the Amerindians during the early settlement period in the West Indies.

## MORINGACEAE

### *Moringa oleifera* Lam.; Horseradish Tree

This tree has alternate pinnately compound leaves with obovate leaflets, showy white to pink flowers, and elongated one-chambered capsules. It is native to India. Specimens were collected at the McChesney Estate in Olveston. *Brussell C-242*.

Tea made by boiling the roots is drunk as an expectorant.

The green pods are cooked and eaten like common

okra. The blossoms, young pods, and young leaves are eaten raw in salads or cooked in stews or vegetable dishes. A horseradish flavored relish is made from the soft roots.

## MUSACEAE

*Heliconia caribaea* Lam.; Balisier; figure 41; plate 20

This perennial herb has alternate, long-stalked, ovate-lanceolate leaves and greenish-yellow flowers protruding from showy yellow or bright red bracts. The seeds are borne in deep blue capsules. It is native to tropical America. Specimens were collected on Katy Hill. *Brussell C-212*.

The crushed leaves are used as poultices, and the sap from the stem is applied topically to treat burns.

The rhizomes are harvested and cooked as vegetables. The leaves are used as a paper substitute to wrap various items such as fish, seeds, fruits, and vegetables and are used for thatching.

*Musa paradisiaca* L.; Plantain

This treelike herbaceous plant has very large, alternate, oblong leaves and clusters of white to purplish-yellow flowers shielded by reddish-purple hoodlike bracts and fleshy, yellow, elongated fruits. It is widespread in the tropics. Plantains were found growing at Paradise. *Sight record*.

Tea made from the leaves is drunk to treat colds.

**Figure 41.** Hut thatched with balisier (*Heliconia caribaea*) and banana (*Musa sapientum*) leaves.

Ripe plantains are eaten fried, baked, or boiled. Green fruits are dried and ground into flour. Plantains are sold in the public market at Plymouth. The leaves are also used for wrapping food and for thatching.

*Musa sapientum* L.; Banana; figures 35 and 41

This treelike herb has cylindrical succulent stems with purple spots and very large, alternate, oblong leaves. The ivory to purplish-yellow flowers are borne in clusters and give rise to fleshy, yellow, elongated fruits. Bananas were found growing at Paradise. *Sight record.*

Bananas are eaten raw by themselves or sliced and mixed in fruit salads. These fruits are also cooked in puddings and baked goods. Bananas are sold in the public market at Plymouth.

The leaves are used to make thatched roofs.

Cooked sweet potato pulp and cooked shredded coconut meat were once mixed with poison and tied up in banana leaves in voodoo practices that involved poisoning someone.

## MYRTACEAE

*Eucalyptus resinifera* J. E. Smith; Eucalyptus

This large tree has reddish-brown peeling bark and alternate, curved, lanceolate leaves that are aromatic when crushed. The cream-colored flowers produce small globose capsules filled with numerous minute seeds. It is native to Australia. Specimens were collected near the Emerald Isle Hotel. *Brussell C-106.*

Tea made by steeping the leaves is drunk to treat colds, asthma, and sinus congestion.

*Eugenia uniflora* L.; Surinam Cherry

This tree has opposite ovate leaves, small white flowers, and glossy, scarlet, eight-ribbed, globose berries. It is native to eastern Brazil. A surinam cherry tree was found growing at Paradise. *Sight record.*

The juicy piquant fruits are eaten raw and made into jellies and preserves.

*Myrcia splendens* (Sw.) DC.; Black Birch, Red Rodwood

This tree has opposite ovate leaves, small white flowers in lateral and terminal clusters, and globose blue drupes. It is native to the West Indies. Specimens were collected at Woodlands. *Brussell C-125.*

The fruits are eaten raw or cooked in pies and preserves. The hard wood is used for posts and tool handles.

*Pimenta racemosa* (Mill.) J. W. Moore; Bay Rum Tree, Celemon Bush, Cinnamon Tree

This tree has opposite elliptic leaves, clusters of fragrant white flowers, and globose, fleshy, black fruits. It is native to the West Indies. Specimens were collected at Salem and Olveston. *Brussell C-298, C-351.*

A tea made by boiling the leaves and fruits is drunk to treat coughs, colds, nausea, and intestinal disorders. Tea made by boiling the leaves is used to cook turtle meat on the island.

This tree is highly valued as a source of myrica oil which is used in the production of bay rum. Researchers on Montserrat have demonstrated that 45 kilograms of bay rum tree leaves yield approximately 580 milliliters of myrica oil (Williams and Williams 1951). Bay rum is made up of myrica oil, orange oil, pimenta oil, water, and alcohol (Little and Wadsworth 1964). Myrica oil is also used in perfumery. In the past, myrica oil was an important export for Montserrat.

### *Psidium guajava* L.; Guava

This small tree has opposite oblong leaves, white flowers, and yellow globose fleshy berries with numerous seeds. It is native to tropical America. Specimens were collected at Salem and Paradise. *Brussell C-75, C-263, C-291.*

The leaves, bark, and roots are boiled to make an astringent tea that is drunk to treat diarrhea. Tea made by boiling the leaves is mixed with donkey milk and drunk to treat colds. According to Duberry (1973) the fresh leaves are chewed as a cold remedy.

The fruits are eaten raw, canned whole, and made into juice, jelly, and guava cheese (a solidified jam or jelly). The fruit is high in iron and has been found to contain two to five times the ascorbic acid content of fresh orange juice (Duberry 1973, Merrill 1954). Guava fruits are also a good source of vitamin A (Merrill 1954).

Montserratians pound one end of a guava twig with rocks and use the fibrous end as a very effective toothbrush.

Guava fruits are sold in the public market at Plymouth for culinary use, and guava leaves are also found there in medicinal "tea bundles" (dried and wrapped bundles of leaves and branches of medicinal plants).

### *Syzygium cumini* (L.) Skeels ( = *Eugenia cumini* [L.] Druce); Java Plum, Jambolan

This tree has alternate ovate leaves and small white flowers that give rise to juicy, purple, plumlike fruits. It is native to tropical Asia. A Java plum tree was found growing at Olveston. *Sight record.*

The fruits are eaten raw and cooked in preserves.

### *Syzygium jambos* (L.) Alston ( = *Eugenia jambos* L.); Rose Plum, Rose Apple

This tree has opposite lanceolate leaves, large showy, white "powderpuff" flowers in terminal clusters, and globose pinkish-yellow berries. It is native to tropical Asia. Specimens were collected at the Tar River Estate. *Brussell C-73.*

The fruits are eaten raw, whole or cut up in salads. They are also cooked in pies and preserves.

*Syzygium malaccense* (L.) Merr. & Perry (= *Eugenia malaccensis* L.); Malay Apple, Rose Apple; figure 42

This tree has opposite oblong leaves, clusters of pink brushlike flowers, and crisp, juicy, pear-shaped, pinkish to dark red fruits. It is native to tropical Asia. Specimens were collected on St. George's Hill. This is the first report of this species from Montserrat. *Brussell C-239.*

The fruits are eaten fresh, whole, or cut up in salads. They are also used to make preserves and wine. The somewhat tart flowers are added to fruit salads.

I spent a sybaritic afternoon under one of these beautiful trees eating the delicious succulent fruits.

Figure 42. *Syzygium malaccense,* Malay apple.

## OLEACEAE

*Jasminum multiflorum* (Burm. F.) Andr.; Jasmine

This semi-erect shrub has opposite ovate leaves, fragrant white flowers borne in terminal and axillary cymes, and didymous purple berries. It is native to India. Specimens were collected at Olveston. *Brussell C-300.*

The flowers and bark chips are soaked overnight

## Potions, Poisons, and Panaceas

in water, and the resulting medicinal liquid is administered as drops for sore eyes (Duberry 1973).

Tea made from the leaves and flowers is a popular beverage.

### PASSIFLORACEAE

*Passiflora edulis* Sims; Passionfruit; figure 43

This climbing vine has unbranched tendrils, alternate ovate leaves, large ornate white and purple flowers, and globose yellow juicy berries. It is native to South America. Specimens were collected at Woodlands. *Brussell C-149*.

An infusion made by pouring boiling water over the flowers is drunk as a somniferant tea. Tea made by boiling the leaves is drunk as a vermifuge. A decoction made by boiling the roots is drunk for emetic purposes.

The delicious aromatic fruits of this vine are eaten raw or made into refreshing drinks. One need only taste the succulent fruit of this plant to develop a passion for it.

*Passiflora foetida* L.; Baby Honeysuckle

This climbing vine has unbranched tendrils and alternate, three-lobed, cordate leaves. The purple and white flowers give rise to yellow ovoid berries. It is native to tropical America. Specimens were collected in the White River Ghaut. *Brussell C-56, C-108, C-304*.

All parts of the plant are boiled to make a tea that

Figure 43. *Passiflora edulis*, passionfruit.

is drunk to treat colds and congestion of the nasal passages and lungs.

The sweet succulent fruits are eaten raw, and the fresh juice is drunk as a refreshing beverage.

## PHYTOLACCACEAE

### *Petiveria alliacea* L.; Batroot, Strong Man's Weed

This perennial suffrutescent herb has alternate elliptic-ovate leaves, spikes of greenish-white flowers, and elongated achenes bearing four hooks. The entire plant has an alliaceous garlic-like redolence. It is native to tropical America. Batroot was found at Cudjoehead. *Sight record.*

The roots are soaked overnight in water, and the resulting infusion is taken internally in small amounts to relieve stomach pain. Tea made by boiling the entire plant is drunk to treat colds and weakness of the bladder. This same tea is taken in stronger doses to induce abortion. A decoction made by boiling the leaves and roots is applied externally to relieve pain. A decoction made from the leaves is used as a deodorant bath.

When cattle graze on this plant, their milk takes on an alliaceous taste and smell.

## PIPERACEAE

### *Peperomia pellucida* (L.) Kth.; Inflammation Bush

This delicate transparent herb has alternate transparent deltoid-ovate leaves, minute flowers borne in terminal spikes, and gummy drupelike ellipsoid ribbed fruits. It is native to the American tropics. Specimens were collected in the Galway Soufriere area and in the Centre Hills along a path. *Brussell C-11, C-90.*

Tea made by boiling the entire plant is given to overactive children as a sedative. This same tea is drunk to break up a cough and for its purgative and laxative effects. The leaves are used as a poultice to draw the "inflammation" (pus) from a boil. After washing, a single leaf is pressed flat on a clean hard surface by rolling the leaf with a clean bottle until the "ribs" (veins) are pressed flat. The flattened leaf is then coated with petroleum jelly and applied to the boil (Duberry 1973).

### *Piper dilatatum* Rich.; Giant Tree, Elder Bush, Rock Bush

This shrub has alternate rhombic-elliptic leaves, tiny greenish-white flowers borne in spikes, and small drupaceous fruits. It is native to the West Indies. Specimens were collected in the Centre Hills. *Brussell C-82.*

Montserratians use the foliage for bodily cleansing by rubbing the leaves over their skin.

## PLANTAGINACEAE

*Plantago major* **L.; English Plantain**

This perennial herb has a rosette of radial oblanceolate leaves, yellowish-brown flowers borne in spikes, and globose fruits containing many small seeds. It is native to the Old World. English plantain was found growing at Olveston. *Sight record.*

A decoction made by boiling the leaves is used as an eye wash and is drunk to treat asthma. The leaves are soaked in sweet oil and used as poultices to draw the pus out of boils and infected wounds.

## POACEAE

*Bambusa vulgaris* **Schrad. ex Wendl.; Bamboo**

This caespitose, giant, woody grass has bright green or yellow culms, linear-lanceolate leaves, and closely flowered spikelets that give rise to yellowish-brown caryopses. It is native to the Old World tropics. Specimens were collected in the Centre Hills. *Brussell C-224.*

A decoction made by boiling all parts of this plant is drunk to treat colds.

Cups, furniture, and fishing poles are made from the large hollow culms.

*Cymbopogon citratus* **(DC. ex Nees) Stapf; Sweet Grass, Lemon Grass, Fever Grass**

This tufted aromatic perennial has linear leaf blades rolled in the bud and racemes of paired spikelets that produce tiny caryopses. It is native to India. Specimens were collected in the Plymouth market and at the Tar River Estate. *Brussell C-63, C-74.*

Tea made by boiling the leaves is drunk as a satisfying beverage as well as to treat colds and fevers.

Bundles of sweet grass are placed in closets to impart a pleasant fresh scent to clothes.

*Cynodon dactylon* **(L.) Pers.; Hardgrass, Devil's Grass, St. Augustine's Grass, Bermuda Grass**

This stoloniferous perennial grass has strong rhizomes, leaves rolled in the bud, and an inflorescence of five terminal digitate spikes that bear ovoid caryopses. It is widespread in the warmer areas of the world. It was collected in an old roadbed at Paradise. *Brussell C-185, C-312.*

Tea made from the entire plant is drunk and applied topically while warm to treat back pain.

This species is often used as a lawn grass in the tropics and is employed for this purpose on Montserrat.

*Eleusine indica* **(L.) Gaertn.; Dutchgrass, Lovegrass**

This tufted annual grass has leaf-blades folded in the bud and an inflorescence of terminal spikes that bear ovoid caryopses. It is native to the Old World. Specimens were collected at Paradise. *Brussell C-184.*

Tea made by boiling the entire plant is drunk as a treatment for menstrual problems.

It is a Montserratian custom to gather Dutch grass, pound it with stones, and mix the pulverized grass with vinegar and turpentine. The resulting liquid is then used to bathe a dog in order to turn the canine into a mean watchdog.

*Saccharum officinarum* L.; Sugarcane; figures 44, 45, and 46; plate 21

This robust, tall tufted perennial grass has leaf blades rolled in the bud, membranaceous ligules, and large panicles bearing yellowish-brown caryopses. Originally from the East Indies, it is widely cultivated in tropical areas. Specimens were collected at Woodlands. *Brussell C-129.*

Tea made by boiling the upper stems and leaves is drunk as a cold treatment.

The inflorescences of this plant were once used to stuff pillows on the island.

Sugarcane was once a very important crop on the island and was the source of sugar, molasses, and rum. The several surviving stone sugar mills stand as impressive monuments to the bygone days when Montserrat was a sugar-producing island. The Montserrat National

**Figure 44.** Abandoned sugar mill built in the early eighteenth century at Trants Estate.

**Figure 45.** Sugarcane crusher at Trants Estate.

**Figure 46.** Iron sugarcane juice evaporating kettle surviving from the days when Montserrat was a sugar-producing island.

Trust has turned one of these sturdy sugar mills into a fascinating museum.

Presently sugarcane is mostly chewed raw or used to make local vinegar. This vinegar is made by cutting the cane into small pieces, then crushing the pieces of cane in water and allowing the mixture to set in a glass jar in the sun for two months. Aside from its culinary uses, this vinegar is rubbed into the scalp to treat headaches and rubbed on the body for treating aches and pains. It is applied gently to the skin to treat sunburn.

An appealing rum punch is produced and bottled on Montserrat for export.

*Trichachne insularis* (L.) Nees ( = *Digitaria insularis* [L.] Mez ex Ekman); Long Grass, Corn Grass

This erect perennial grass has flat linear leaves and narrow panicles of acuminate spikelets that bear chestnut-brown caryopses. It is native to tropical America. Specimens were collected in the White River Ghaut, at Woodlands, and at the roadside near Galway Soufriere. *Brussell C-8, C-57, C-124.*

Tea brewed from the roots and culms is drunk to treat muscle strains and back pain.

The leaves are used to clean teeth. The leaf blades are folded longitudinally and rubbed over the surface of the teeth and drawn back and forth between adjacent teeth in the same way that dental floss is used. When the crushed leaves are used to scrub the teeth, the bitter juice that results is said to freshen the breath and prevent tooth decay.

This plant is used as food for cattle, rabbits, pigs, goats, and donkeys. Montserratians believe that this grass stimulates milk production in cattle.

*Vetiveria zizanioides* (L.) Nash; Vete Vere Grass, Khus Khus, Sweet Grass

This tufted perennial grass has strongly keeled linear leaf blades rolled in the bud and panicles of spiny spikelets that bear tan caryopses. It is native to Asia. Specimens were collected at Olveston. *Brussell C-95.*

The aromatic roots of this grass are put in closets and chests to make clothes smell fragrant. The leaves and culms are used for weaving mats, which are sold in the public market at Plymouth.

*Zea mays* L.; Corn

This tall grass has linear-lanceolate leaves, a terminal tassel bearing staminate flowers, and cobs bearing pistillate flowers, borne lower on the culm. The mature grains are produced in rows on the cobs. Corn is native to tropical America. It was seen growing at Harris Village. *Sight record.*

For the treatment of jaundice, Montserratians parch corn until it is black, grind it, and brew a tea from the grindings that is administered to the patient in one cup doses four times a day.

Fresh corn on the cob is eaten roasted or boiled and is sold in the public market at Plymouth.

## POLYGONACEAE

*Antigonon leptopus* Hook. and Arn.; Coralita

This herbaceous climbing vine has alternate ovate leaves, pendulous racemes of showy pink flowers, and black achenes that are each enclosed in an accrescent calyx. It is native to Mexico. Specimens were collected at the Yacht Club Beach and at the Groves in Plymouth. *Brussell C-5, C-372.*

The leaves and flowers of this beautiful climbing vine are boiled to make a tea that is drunk to treat colds and congestion of the sinuses and lungs. I spoke with people on both Montserrat and Dominica who highly esteem coralita as a cold treatment and later used coralita tea myself to successfully treat the symptoms of sinus congestion.

*Coccoloba swartzii* Meissn.; Red Grape

This tree has alternate ovate leaves, small greenish-white flowers, and three-angled nuts that are each enclosed by the persistent fleshy perianth. It is native to the West Indies. Red grape trees were found growing near the sea, east of the Tar River Estate. *Sight record.*

The hard heavy wood is used to make tool handles. Exudate from the root bark is used to tan hides.

*Coccoloba uvifera* (L.) L.; Seagrape; plate 22

This tree has alternate orbicular leaves, white flowers in dense spike-like racemes, and purplish-green three-angled nuts that are each enclosed by the persistent fleshy perianth. It is native to the West Indies. Specimens were collected near the Radio Antilles building in the White River Valley. *Brussell C-280.*

A decoction made by boiling both the green and mature fruits is drunk to treat dysentery and venereal disease.

The ripe fruits are eaten raw and are also used to make beverages and jelly.

*Coccoloba venosa* L.; Chiddle Grape

This tree has alternate obovate leaves, racemes of yellowish-green flowers, and three-angled nuts that are each enclosed in a succulent, white, persistent fleshy perianth. It is native to the West Indies. Chiddle grape trees were found growing near the Hot Water Pond. *Sight record.*

Tea made from the leaves, flowers, and fruits is drunk to treat colds and stomach pains.

The sweet succulent calyces are eaten raw or used to make preserves.

## POLYPODIACEAE

*Polypodium heterophyllum* L.; Climbing Tongue Fern, Snake Bush

This slender creeping fern has alternate linear-lanceolate leaves with one row of sori on each side of the midvein. It is native to the West Indies. Specimens were collected in the Centre Hills in a wet forest. *Brussell C-79.*

The entire plant is boiled, and the resulting decoction is drunk to treat colds and back pain.

## PORTULACACEAE

*Portulaca oleracea* L.; Pusly, Pursly, Purslane

This prostrate succulent herb has alternate obovate leaves, yellow flowers, and circumcisal capsules containing many seeds. It is widely distributed in tropical and warm temperate regions. Previously this species has generally been regarded by taxonomists as a nonnative historical introduction to the New World. However, carbonized seeds of *P. oleracea* found in a late woodland feature at the Carrier Mills archaeological site in southern Illinois provide evidence for the presence of this species in the New World during pre-Columbian times (Lopinot and Brussell 1982). Specimens were collected at the Tar River Estate and in the White River Valley. *Brussell C-71, C-101.*

A decoction made by boiling the entire plant is drunk as an anthelmintic.

The leaves and stems are eaten raw in salads or cooked as potherbs and eaten like spinach (cf. Hodge and Taylor 1957).

## PROTEACEAE

*Macadamia ternifolia* F. Muell.; Macadamia

This tree has whorled oblong leaves, white flowers in drooping clusters, and globose fruits that each contain a single whitish edible seed. It is native to Queensland, Australia. Specimens were collected at the McChesney Estate in Olveston. *Brussell C-246.*

The seeds are roasted and eaten.

These seeds bring a very high price on world markets. The trees usually begin to bear fruit after the fourth or fifth year. By the eighth year, some trees have been known to bear eight hundred pounds of seeds per year. Due to the high market value, ease of storage and shipping, and suitability for the island's climate and topography, the production of macadamia nuts seems to have great potential as a cash crop that would stimulate the economy of Montserrat.

## PSILOTACEAE

*Psilotum nudum* (L.) Griseb.; Tickleweed

This terrestrial or epiphytic herbaceous plant has a dichotomously forked stem, alternate prophylls, three-chambered synangia, and mealy oval spores. It is widespread in the tropical areas of the world. Specimens were collected at the base of Chance Peak growing on *Mangifera indica* L. *Brussell C-219.*

A decoction made by boiling the entire plant is drunk as a laxative. The spores are used as baby powder.

## PUNICACEAE

***Punica granatum*** **L.; Pomegranate, Granada**

This small tree has opposite oblong-lanceolate leaves, showy scarlet flowers, and globose red fruits filled with pink juice sacs that each contain one white triangular seed. It is native to southern Asia. Specimens were collected at Olveston. *Brussell C-241.*

The root bark, fruit rind, and twigs are boiled to make a decoction that is drunk to treat dysentery. Tea made by boiling the fruit rind is drunk as a mild sedative and a "cure for belly-ache."

The fruits are eaten raw and used to make syrup and refreshing drinks.

## RHAMNACEAE

***Colubrina elliptica*** **(Sw.) Briz. & W. Stern; Mawby, Soldierwood**

This tree has alternate elliptic leaves, clusters of tiny yellowish-green flowers, and globose three-lobed orange fruits that contain shiny dark brown seeds. It is native to the West Indies. Mawby trees were found growing near Harris Village. *Sight record.*

A decoction made by boiling the roots, bark, and leaves is drunk as a purgative.

Mawby, a popular fermented drink, is produced from the bark of this tree. The bark is boiled with cloves and molasses. During the boiling process, the froth and impurities are skimmed off the surface of the decoction. After cooling, the liquid is left to ferment for one day and skimmed once again. The beverage is then put in uncapped bottles to avoid explosions from gas build up due to continued fermentation. Mawby is said to "build the system"; in other words it acts as a tonic. It is also drunk as a refreshing beverage that facilitates digestion and stops diarrhea. I successfully used this sweet pungent libation to treat a case of indigestion resulting from eating too much breadfruit au gratin. Some people drink this concoction as a potion that is believed to induce amorous revelry.

## ROSACEAE

***Rubus rosifolius*** **J. E. Smith; Raspberry, Fraise**

This prickly shrub has recurved canes, alternate compound leaves with elliptic leaflets, white flowers, and reddish-purple fruits. It is native to tropical America. Raspberry canes were found in a hillside thicket at Paradise. *Brussell C-187.*

Tea made by boiling the leaves and roots is drunk to stop diarrhea. The fresh fruits are eaten for their cleansing laxative effect.

The fruits are eaten raw and cooked in pies, jelly, and preserves.

## RUBIACEAE

*Chiococca alba* (L.) Hitchc.; David's Root, Davis Root, Briny Roots

This scrambling shrub or climbing vine has opposite ovate leaves, small white flowers in racemes, and white globose drupes. It is native to the West Indies. Specimens were collected at Centre Hills, White River Ghaut, Salem, and Great Alp Falls. *Brussell C-40, C-77, C-270, C-289.*

Tea made by boiling the roots is drunk as a purgative.

Cut up pieces of the aromatic roots are soaked for several days in an uncapped bottle of water to produce a naturally carbonated, fermented, "aphrodisiac" drink that is said to "give vigor and vitality." I had my doubts about the validity of the claims that were made about this plant's qualities; however, the David's root drink that Mr. Fred Payne of Salem prepared had a definite stimulatory effect as well as powerful diuretic properties. This potion also contained whole raw peanuts (*Arachis hypogaea* L.) and chopped pieces of ramgoat bush (*Eryngium foetidum* L.).

This plant gives off a sweet pungent redolence that allowed Fred Payne and me to "sniff it out" in the dense jungle. We literally followed our noses to this fascinating vine. This was my first experience of locating a rainforest plant by the olfactory method.

The root is said to be a "slow poison" that causes itching hives when gathered and used during a full moon.

*Coffea arabica* L.; Coffee Tree

This small tree has opposite elliptic leaves, fragrant white flowers, and globose red berries containing one or two seeds each. It is native to Ethiopia. Coffee trees were found growing in an upper level garden above Salem. *Sight record.*

The raw seeds are eaten to treat arthritis. A decoction made by boiling the fresh leaves is drunk to relieve toothache symptoms and is used as a topical bath to treat body pains.

This small tree is occasionally cultivated on the island for the seeds, which are cured, roasted, and ground to produce the ubiquitous strong, dark beverage (coffee) that is known for its stimulant properties.

*Genipa americana* L.; Genip

This tree has opposite obovate leaves, yellow flowers in terminal clusters, and large elliptic yellowish-brown berries that each contain many flat yellow seeds. It is native to the West Indies. Genip trees were found growing at Woodlands. *Sight record.*

Juice squeezed from the fruits is mixed with water to make a sour refreshing drink that is sometimes fermented and drunk as an intoxicating beverage. The pulp of the fruits is eaten fresh and used to make marmalade. The tough elastic wood is used as lumber for general construction. A dark blue dye is obtained from the unripe fruits.

### *Ixora ferrea* (Jacq.) Benth.; Black Candlewood

This tree has opposite narrowly elliptic leaves, white and red flowers in lateral clusters, and globose pink berries. It is native to the West Indies. Black Candlewood trees were found growing in the Centre Hills. *Sight record.*

The leaves and twigs are boiled, and the resulting decoction is drunk to treat fevers, coughs, and colds.

The wood is used for tool handles, posts, and charcoal.

### *Morinda citrifolia* L.; Hog Apple, Painkiller, Chiddle Grape

This small tree has opposite elliptic leaves, white tubular flowers in balls, and large whitish-green multiple fruits that have an odor similar to limburger cheese. It is native to India. Specimens were collected on St. George's Hill and at Blackburn Airport near Trants. *Brussell C-269, C-311.*

The crushed or heat-wilted leaves are used topically to relieve pain and as poultices for boils, bruises, and wounds. The immature macerated fruits are mixed with salt, and the resulting paste is applied topically to areas around broken bones. Juice from the root is applied externally to treat skin eruptions. Tea made from the leaves and bark is drunk as a tonic.

The odiferous fruits are edible.

### *Randia aculeata* L.; Goat Hahn, Christmas Bush

This small spiny tree has opposite or clustered spatulate leaves, fragrant white flowers, and whitish berries with navy blue pulp and several seeds. It is native to the West Indies. Goat hahn was found at Salem. *Brussell C-171.*

This plant is sometimes used as a Christmas tree on the island. The fruits are the source of a navy blue dye.

## RUTACEAE

### *Amyris elemifera* L.; Torchwood

This small tree has opposite compound leaves with lanceolate leaflets. The small white flowers are borne in corymbose panicles and produce small, black, globose drupes. It is native to the West Indies. Torchwood was found at Woodlands. *Sight record.*

The aromatic resin produced by this tree is burned as incense in religious and voodoo ceremonies. Torches made from the resinous wood are used for hunting crabs and other animals at night (Pulsipher 1977). The wood is used for furniture and fence posts.

## Ethnobotanical Uses and Specific Discussion

***Citrus aurantifolia*** **(Christm.) Swingle; Lime**

This small tree has spines in the leaf axils, alternate elliptic leaves, and white flowers that give rise to green subglobose fruits. It is native to the East Indies. Specimens were collected in a dooryard garden at Woodlands. *Brussell C-228.*

Tea brewed from the leaves is taken to treat sore throats and colds. Sore throats are also treated by drinking lime juice. Boils and skin ulcers are treated by applying fresh lime peel or roasted lime pulp to the affected area. Lime juice is often mixed with bush medicines to make them more effective (Duberry 1973). The macerated berries of *Solanum ciliatum* Lam. are mixed with lime juice, and the resulting concoction is applied to ringworm as an effective treatment. Lime juice is mixed with bicarbonate of soda and taken as a remedy for stomachache and flatulence. Sprains are treated by rubbing fresh lime "skins" over the afflicted area. Lime peel is also used to remove paint and dead skin as well as to keep the neck of a kerosene lamp cool at night. Limeade or tea made from the leaves is drunk to "calm the sex drive."

During the nineteenth century, limes were a major crop on the island, and lime juice was exported in large quantities. The museum of the Montserrat National Trust sells postcards which are reproductions of nineteenth-century advertisements for Montserrat Lime Juice.

***Citrus aurantium*** **L.; Swivel Sweet, Sour Orange**

This tree has alternate ovate leaves, very fragrant white flowers, and globose orange tart fruits. It is native to southeastern Asia. Specimens were collected in an upper level garden above Salem. *Brussell C-80.*

The chopped fresh rind is steeped, and the resulting tea is drunk to treat colds. The dried rind is boiled, and the resulting decoction is drunk as a mild sedative and as a cure for stomachache. Tea made from the leaves is drunk to induce perspiration. The juice is applied topically to stop bleeding from cuts and as an antiseptic.

The rind is eaten fresh, candied, and cooked with the pulp to make marmalade. Sour orange fruits are sometimes sold in the public market at Plymouth.

***Citrus reticulata*** **Blanco; Tangerine, Mandarin**

This tree has alternate ovate leaves, fragrant small white flowers, and orange subglobose sweet loose skinned fruits. It is native to southeastern Asia. Specimens were collected in the Centre Hills. *Brussell C-78.*

Tea made by steeping the leaves is drunk to treat colds and fevers.

The sweet juicy fruits are eaten raw and are sold in the public market at Plymouth.

***Fortunella margarita*** **(Lour.) Swingle; Kumquat**

This small tree has alternate lanceolate leaves, small aromatic flowers with white petals, and yellow-orange oblong fruits. It is native to southeastern China. A kumquat tree was found growing at Olveston. *Sight record.*

The chopped fruits are steeped to make a tea that is drunk to treat colds.

The sweet rind is eaten raw, and the entire fruits are cooked to make preserves and marmalade.

*Triphasia trifolia* (Burm. f.) P. Wils.; Sweet Lime

This shrub has paired spines at the leaf nodes, alternate leaves composed of three ovate leaflets, white fragrant flowers, and elliptic red fruits. It is native to southeastern Asia. Specimens were collected in the White River Valley. *Brussell C-100.*

Tea made by boiling the leaves and fruits is drunk as a refreshing beverage and as a treatment for colds. The tart edible fruits have a sweet limelike taste.

*Zanthoxylum flavum* Vahl; Satinwood

This small tree has alternate odd-pinnate leaves with opposite ovate leaflets. The small yellow flowers produce globose follicles that contain tiny shiny black seeds. It is native to the West Indies. A satinwood tree was found in the Centre Hills. *Sight record.*

An infusion made from the yellow inner bark is used as a wash for treating sore eyes and pink eye.

The bark is also used as a dye for coloring clothing.

*Zanthoxylum monophyllum* (Lam.) P. Wils.; Yellow Prickle, Yellow Hercules, Yellow Harklis; figure 47

This small tree has prickly bark, alternate elliptic leaves, panicles of small white flowers, and subglobose fruits that each burst to free one shiny black seed. It is native to the West Indies. Specimens were collected in the Centre Hills. *Brussell C-84.*

A decoction made by boiling the bark is administered as drops to treat eye infections and sore eyes.

## SAPINDACEAE

*Blighia sapida* König; Akee; figure 48

This small tree has alternate pinnate leaves composed of elliptic leaflets, white flowers, and showy red three-lobed fleshy capsules with three shiny, black, globose seeds in each fruit. Each seed is attached to a creamy white fleshy aril. It is native to West Africa. Specimens were found on the golf course near the Belham River and in a dooryard below Bugby Hole. *Brussell C-203.*

The fresh fleshy cream colored arils of the akee are eaten cooked, provided the fruits are ripe and have split open naturally. Unripe arils are very toxic. It is generally believed that the overripe arils are toxic also. Many Montserratians refrain from eating this fruit due to the danger involved in eating improperly picked or prepared portions. Symptoms of akee poisoning include cyclic thirst-producing vomiting, muscular and

**Figure 47.** *Zanthoxylum monophyllum*, yellow Hercules.

**Figure 48.** *Blighia sapida*, akee.

nervous exhaustion, prostration, unconsciousness, and death (Morton 1982).

The exocarp (rind) of the fruit is used to poison fish.

*Melicoccus bijugatus* Jacq.; Spanish Lime, Genip

This tree has alternate pinnate leaves composed of elliptic leaflets, fragrant greenish-white flowers, and globose green drupes. It is native to northern South America. Spanish lime trees were found at Woodlands. *Brussell C-114, C-143.*

A decoction made by boiling the leaves is drunk to stop coughing and to reduce fevers. The raw fruits are eaten as a treatment for diarrhea.

The fruits have thin, pink, gelatinous pulp that tastes similar to Concord grapes. Raw whole fruits are chewed and sucked on for their flavor. The fruits are also used to make jelly and wine.

It may be hazardous to eat large quantities of Spanish lime fruits. "It is said that the fine fibers of the pulp have caused the death of children, when swallowed, by forming a coating over the lining of the stomach" (Fawcett and Rendle 1910).

## SAPOTACEAE

*Manilkara zapota* (L.) P. Royen; Sapodilla

This large tree has alternate, often clustered, lanceolate-oblong leaves, light green campanulate flowers, and yellowish-brown globose berries with sweet brown flesh and white exudate. It is native to southern Mexico. Sapodilla trees were found at Olveston. *Sight record.*

The peeled ripe fruits are eaten raw or cooked in preserves or may be cooked and candied. The condensed white latex of this tree was in the past the major source of chewing gum. The raw latex is chewed by some Montserratians.

*Mimusops coriacea* (A. DC.) Miq.; Aphrodite's Apple, Passion Apple, Yellow Sapote; plate 23

This medium-sized tree has alternate elliptic to obovate-elliptic leathery leaves. The small whitish flowers are borne in the leaf axils and give rise to globose yellow fruits about 4 cm in diameter that contain three to five shiny brown seeds. It is native to the Mascarene Islands in the Indian Ocean. *M. coriacea* was collected in a mesic forest at Paradise. This species has not heretofore been reported from Montserrat. *Brussell C-268.*

The sweet pulp of the fruit is eaten as a food source. One informant said the fruits are eaten for aphrodisiac purposes.

## SCROPHULARIACEAE

*Digitalis purpurea* L.; Heart Bush

This pubescent biennial herb has alternate oblong to oblanceolate leaves, racemes of tubular purple to white flowers, and septicidal capsules bearing numerous seeds. It is native to Europe. Specimens were collected at Woodlands. *Brussell C-120.*

The leaves are boiled, and the resulting decoction is drunk for heart trouble. This preparation is said to slow a rapid heart.

## SIMAROUBACEAE

*Picrasma antillana* (Eggers) Urb.; Bitter Bark, Bitter Ash

This tree has alternate pinnately compound leaves composed of narrowly elliptic leaflets, small yellowish-green flowers in corymbose axillary panicles, and reddish-black drupes. It is native to the West Indies. Specimens were collected in a forest above Salem. *Brussell C-296.*

An infusion prepared by soaking the wood chips overnight in water is drunk as a treatment for high blood pressure, diabetes, dyspepsia, and as a febrifuge. Cups carved from the wood are called "bitter cups," and when water is allowed to stand in one of these cups, the same bitter infusion results. This "bitter cup water" is drunk for the same purposes as the wood chip infusion. Tea made by boiling the bark of this tree is mixed with *Momordica charantia* leaf tea, and the resulting preparation is drunk to treat cancer. (cf. Hodge and Taylor 1957)

*Simarouba amara* Aubl.; Tom Tah, White Deal, Snakewood

This tree has alternate pinnately compound leaves composed of oblong leaflets, small greenish-white flowers, and ellipsoid drupelets. It is native to the West Indies. Specimens were collected in a mesic forest above Salem. *Brussell C-168.*

The wood from this important lumber tree is used for general construction.

## SMILACACEAE

*Smilax cumanensis* Humb. & Bonpl. ex Willd.; Wild Sarsaparilla

This woody vine has alternate large oval leaves, small greenish flowers in umbellate clusters, and dark reddish-blue berries. It is native to the West Indies. Specimens were collected at Paradise. *Brussell C-267.*

The roots are boiled, and the resulting decoction is drunk for relieving muscle strains.

The fresh crushed roots are put in containers of water, and the mixture is allowed to ferment and then drunk as beer. This sarsaparilla beer is said to have aphrodisiac properties and is sometimes mixed with a *Chiococca alba* decoction for a heightened effect.

## SOLANACEAE

*Capsicum annuum* L.; Hotpepper Bush

This suffrutescent herb has alternate narrowly ovate leaves, white flowers, and red or yellow ovoid berries. It is native to tropical America. Specimens were collected at Woodlands in a dooryard garden. *Brussell C-111.*

The fruits are chewed raw to relieve sinus congestion.

Hotpepper fruits are used for seasoning cookery and are also pickled and eaten as condiments.

*Datura innoxia* Mill.; Wonga, Painkiller; figure 49

This suffrutescent herb has alternate ovate leaves, white flowers, and globose prickly four-valved capsules with numerous seeds. It is native to tropical America. Specimens were collected at the Trants Estate. *Brussell C-51, C-196.*

*Datura innoxia* is a well-known anaesthetic and hallucinogen. It was said to have been the drug plant that was most widely used by Indians of the American Southwest. The alkaloid composition of *D. innoxia* is quite similar to that of *Datura metel* L. (Schultes and Hofmann 1980).

The fresh leaves are heated and applied topically on swollen joints and wounds as pain-relieving poultices.

The seeds are especially known for their halluci-

**Figure 49.** *Datura innoxia*, wonga.

nogenic properties and have been used in voodoo practices on the island.

Plants of the *Datura* genus are poisonous; therefore, I do not recommend using any part of these plants.

### *Datura metel* L.; Wanger Bush

This suffrutescent herb has alternate ovate leaves, white or purple-tinged flowers, and ovoid prickly four-valved capsules containing numerous seeds. It is native to the Old World tropics. *D. metel* was found at Woodlands. *Sight record.*

The crushed and heated fresh leaves are placed on sore joints and bruises to relieve pain and swelling. The crushed and heated leaves are also placed on tumors to remove them.

*Datura* has a long history of use as a hallucinogen and stuporific poison. *Datura stramonium* L. is one of the ingredients employed in the Haitian zombification process (Davis 1985, 1988).

On one occasion while harvesting soybeans on our family farm in Cumberland County, Illinois, my father accidently got a *Datura stramonium* seed in his eye. Even though the seed was removed in less than ninety seconds, his pupil became greatly dilated and remained that way for twenty-four hours.

Scopolamine, norscopolamine, meteloidine, hyoscyamine, norhyoscyamine, nicotine, and cuscohygrine are the alkaloids that have been found in *D. metel*. The total alkaloid contents were determined to be: 0.2 to 0.5 percent in the seeds, 0.1 to 0.2 percent in the roots, 0.2 to 0.5 percent in the leaves and 0.12 percent in the fruits (Schultes and Hofmann 1980).

### *Datura suaveolens* Humb. & Bonpl.; Angel's Trumpet; plate 24

This arborescent shrub has alternate ovate-oblong leaves, very large white flowers, and four-valved capsules containing many seeds. It is native to tropical America. *D. suaveolens* was found at Salem. *Sight record.*

The leaves are heated on a hot rock and placed on aching feet and sore joints to relieve pain.

The seeds have been mixed with food to poison enemies. Scopolamine is the major active alkaloid that is found in all species of the *Datura* genus (Schultes and Hoffman 1980).

### *Lycopersicon esculentum* Mill.; Tomato

This herb has alternate compound leaves composed of irregularly lobed leaflets, yellow flowers, and globose to pyriform red or yellowish-orange berries. It is native to the Andes Mountains. Tomatoes were found growing at Harris Village. *Sight record.*

Tomatoes are commonly eaten both fresh and cooked and are used to make tomato marmalade. Tomato fruits are sold in the public market at Plymouth.

***Nicotiana tabacum* L.; Creole Tobacco**

This tall herb has alternate oblong-lanceolate leaves, large pink funnel-shaped flowers, and ovoid two-chambered septicidal capsules with many small seeds. It is native to South America. Specimens were collected at St. George's Hill in a dooryard garden. *Brussell C-316.*

Tea made by boiling the leaves is drunk as a sedative and cold treatment. The fresh or dried leaves are used as a healing bandage for wounds and blisters. The heated leaves are applied externally to relieve pain from wounds and bruises. Tobacco leaves are chewed to treat toothache. The dried leaves are smoked for their antispasmodic and sedative properties.

***Solanum ciliatum* Lam.; Jumbie Tomato**

This shrub has alternate ovate yellow prickle-bearing leaves, white flowers, and yellow or orange berries. It is widespread in tropical areas of the world. Jumbie tomatoes were found at Salem. *Sight record.*

Juice from the macerated leaves is applied topically to treat ringworm.

***Solanum ficifolium* Ort.; Pigeonberry, Red Trouble, Red Trubba**

This shrub has alternate ovate leaves bearing prickles, white flowers, and small red globose berries. It is widespread in tropical areas of the world. This plant was found growing in the Galway Soufriere area. *Brussell C-12.*

A decoction made by boiling the fruits is drunk to expel intestinal worms and to improve impaired vision.

***Solanum melongena* L.; Eggplant, Boulangie; figure 50**

This suffrutescent herb has alternate ovate leaves, purple flowers, and purple or yellow pyriform berries. It is native to Asia. Eggplants were found growing at Trants. *Sight record.*

A decoction made by boiling the roots is drunk as a purgative.

The fruits are eaten baked or fried. Eggplant fruits are sold in the public market at Plymouth.

## STERCULIACEAE

***Theobroma cacao* L.; Cocoa Tree, Cacao; figures 51 and 52**

This small tree has alternate ovate leaves, small cauliferous white flowers, and large elliptic-ovoid five-chambered reddish-purple and yellow ribbed fruits containing numerous, oily, ovoid seeds. It is native to South America. Specimens were collected in an upper level garden above Salem. *Brussell C-27, C-290.*

Poultices made of the crushed leaves are applied to wounds to speed the healing process and to prevent infection.

**Figure 50.** Informant with *Solanum melongena*, white eggplant.

**Figure 51.** Fruit of *Theobroma cacao*, cocoa.

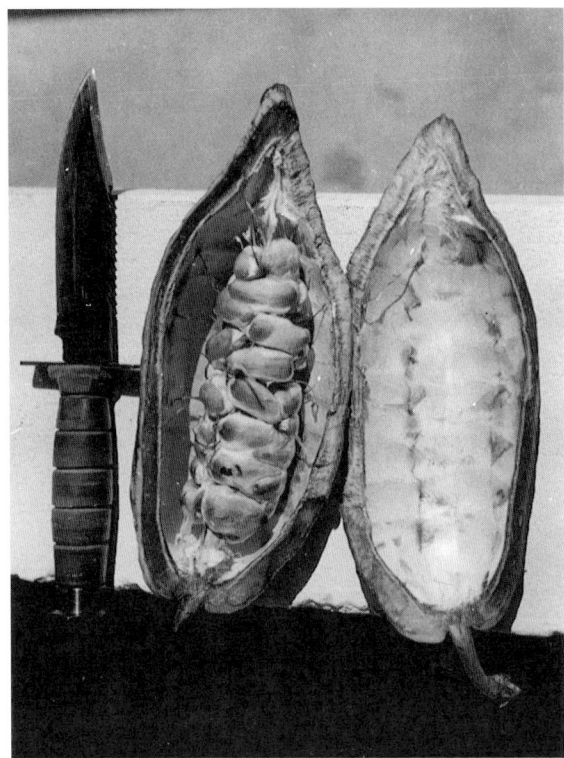

**Figure 52.** Opened fruit of *Theobroma cacao*, cocoa, showing the numerous seeds. Raw cocoa powder obtained from the seeds is used to make a highly stimulating local beverage.

The seeds are fermented, roasted, and ground. The resulting powder is made into balls using sweet oil as a vehicle. These cocoa balls are dropped into boiling water to make a very strong stimulant drink. I found this drink to have a much greater stimulant effect than black coffee. Cocoa balls are sold in the public market at Plymouth.

The sweet tart pulp that is found within the fruits is eaten fresh by Montserratians and is a special favorite of children.

## THEOPHRASTACEAE

### *Jacquinia armillaris* L.; Currant Tree

This small tree has whorled obovate leaves, white funnelform flowers, and globose orange berries. It is native to the West Indies. Currant trees were found growing at Olveston. *J. armillaris* was cited by Shafer (1907). *Sight record.*

The crushed fruits are used to stupefy fish.

### *Jacquinia berterii* Spreng.; Bois Bande

This small tree has opposite or whorled spoon-shaped leaves, small yellow flowers, and egg-shaped orange berries. It is native to the West Indies. Bois bande was found growing at Silver Hill. *Sight record.*

Tea made by boiling the leaves and bark is drunk as a nerve stimulant.

## TROPAEOLACEAE

*Tropaeolum tuberosum* **R. & P.; Biscuit Plant**

This twining tuber-bearing herb has alternate peltate leaves, showy yellow-orange flowers, and three-lobed schizocarps. It is native to South America. Specimens were collected at St. George's Hill. *Brussell C-320.*

Tea is made from the leaves of this plant. The tubers are cooked and eaten like white potatoes. The fresh stems, fruits, and seeds are pickled and used in relishes.

## URTICACEAE

*Fleurya aestuans* (L.) **Miq.; Broad Leaf Nettle, Nettle**

This erect herb is covered with stinging hairs and has alternate ovate leaves and small inconspicuous greenish flowers that give rise to tiny oblique achenes. It is widespread in tropical areas of the world. Specimens were collected on a mesic slope above Salem. *Brussell C-179.*

Tea made by boiling the leaves is drunk for "stoppage of water" (inability to pass urine) and to treat colds.

## VERBENACEAE

*Citharexylum spinosum* **L.; Fiddlewood**

This tree has opposite ovate leaves, white fragrant flowers in racemes, and red to black drupes. It is native to the West Indies. Specimens were collected at St. George's Hill. *Brussell C-317.*

A decoction made by boiling the leaves of this tree and the fruits of *Capsicum frutescens* L. is drunk to treat shortness of breath.

The wood is used for general construction, musical instruments, and furniture.

*Lantana camara* **L.; Lantana, Red Sage**

This shrub has four-angled subcylindrical stems with or without prickles, opposite ovate leaves, red or yellow-orange flowers, and shiny black drupes. It is native to the West Indies. Specimens were collected at Cudjoehead Village and in the Centre Hills. *Brussell C-88, C-337.*

The leaves are boiled to make a decoction that is drunk to treat colds, coughs, arthritis, and dyspepsia.

Goats, cattle, and donkeys are poisoned by eating the leaves.

*Lantana reticulata* **Pers.; Measle Bush**

This pubescent shrub has four-angled subcylindrical stems, opposite ovate leaves, small white flowers, and wrinkled black drupes. It is native to the West Indies. Specimens were collected on a slope above Salem. *Brussell C-164.*

Tea made by boiling the leaves and stems is drunk to treat colds and measles.

***Lippia micromera*** Schau.; Balsam, Oregano

This diffuse shrub has opposite elliptic leaves, fragrant white flowers, and tiny nutlets. It is native to northern South America. *L. micromera* was found growing at the south edge of Plymouth. *Sight record.*

The leaves and stems are boiled to make a decoction that is drunk to treat colds and sore throats.

The dried and powdered leaves are used to season cookery in the same way that commercial oregano is utilized. Dried bundles of this plant were found in the public market at Plymouth.

***Lippia nodiflora*** (L.) Michx.; Man Better Man

This low creeping perennial herb is densely puberulent and has opposite spatulate leaves, small purple to white flowers, and sharply pointed nutlets. It is found in warm temperate and tropical areas of the Old and New World. Man better man was collected in the White River Valley. *Brussell C-278.*

Tea made by boiling the leaves and stems is drunk to treat colds and for general internal "cleaning" of the body.

***Stachytarpheta cayennensis*** (Rich.) Vahl; Worwine, Whorewine, Devil's Tail, Eyebright

This suffrutescent herb has opposite ovate leaves, pale blue to white flowers, and linear one-seeded nutlets. It is native to the West Indies. Specimens were collected at Salem, Woodlands, and Paradise. *Brussell C-17, C-139, C-181.*

The leaves are boiled to make a tea that is given to teething babies. Fresh leaves are placed on skin ulcerations to "draw out poisons," remove pus, and accelerate healing. Drops of juice squeezed from the stems are applied topically to "clear the eyes."

***Stachytarpheta jamaicensis*** (L.) Vahl; Eyebright, Blue Flower, Vervain

This annual herb has opposite ovate leaves, blue flowers, and linear one-seeded nutlets. It is native to the West Indies. Specimens were collected at Salem. *Brussell C-287.*

The stems and leaves are crushed and small drops of the resulting juice are applied to the eyes to clear them. Tea made by boiling the leaves is drunk to treat stomach trouble, fevers, and high blood pressure.

***Tectona grandis*** L.; Teak

This tree has opposite ovate leaves, numerous small white flowers, and finely pubescent brown drupes. It is native to southern Asia. A teak tree was found in an estate garden at Olveston. *Sight record.*

The lumber from this valuable timber tree is used for general construction.

## VITACEAE

*Cissus sicyoides* L.; Chorita, Skip Rope, Masquerade Whip, Poison Wythe, Scratch Wythe

This woody climbing vine has alternate cordate leaves, purple or yellow flowers in corymbose cymes, and black globose berries. It is native to the West Indies. Specimens were collected at Cork Hill, Galway Soufriere, Woodlands, and Paradise. *Brussell C-25, C-31, C-193, C-374.*

A decoction made by boiling the bark is drunk to treat arthritis and backache.

Poultices made from the crushed fresh leaves are applied to painful sores and wounds. This application is said to relieve the pain as well as to speed healing.

Children use portions of this vine for "skipping ropes."

## ZINGIBERACEAE

*Alpinia zerumbet* (Persoon) Burtt & R. R. Smith (= *Alpinia nutans* [Andres] Roscoe); Shell Plant, Shell Ginger

This tall herb has large, alternate, lanceolate-oblong leaves and showy, fragrant, pink and white yellow-tinged flowers in terminal racemes. The pulpy loculicidal capsules bear many small seeds. It is native to China. Specimens were collected in a dooryard garden at Salem. *Brussell C-167.*

Tea made by boiling the leaves and flowers is drunk to treat indigestion, muscular strains, and arthritis.

Beer is made from the roots. The roots are also used as a flavoring agent to replace ginger in cookery.

*Zingiber officinale* Roscoe; Ginger

This perennial herb has alternate linear-lanceolate leaves and red, yellow, and blue flowers that are borne in short dense spikes. The pulpy capsules contain many small seeds. It is native to tropical India. Specimens were collected in an upper level garden at Woodlands. *Brussell C-226.*

The rhizomes are cut into small pieces and boiled. The resulting decoction is allowed to cool and is then drunk as an effective remedy for gastric distress. A few pieces of ginger rhizome are added to beans while they are cooking in order to prevent flatulence in those who eat the beans.

Ginger beer, a popular drink on Montserrat, is made by allowing fermentation to occur in a sugar-water solution that contains chopped pieces of ginger rhizome. Ginger rhizome is candied and is also used as a flavoring agent in various dishes prepared by Montserratians.

## ZYGOPHYLLACEAE

*Guaiacum officinale* L.; Lignum Vitae, Strongwood

This tree has opposite, abruptly pinnate, compound leaves composed of elliptic leaflets and blue or white flowers that produce yellow, obcordate, compressed fruits with ellipsoid seeds and white arils. It is native to the West Indies. Specimens were collected at Shoe Rock. *Brussell C-256, C-257.*

The resin that exudes from this tree is applied topically to treat arthritis, sore muscles, and ringworm. Tea made by boiling the leaves is drunk as a cold medicine, diaphoretic, stimulant, rheumatism remedy, and abortifacient. Some informants state that the leaves are poisonous. It is known that an overdose of lignum vitae tea may cause acute toxic nephritis (Bayley 1949).

The very hard strong wood is self-lubricating due to the fact that it contains an oily resin that is released by the heat of friction. This valuable wood is used for pulley blocks, bushings, bearings, and propeller drive shafts of ships as well as for mallets and mortars and pestles.

**Appendix**

**References Cited**

**Index of Common Names**

**Index of Scientific Names**

# Appendix: A Plant Collection from Montserrat

The following list consists of plants collected during my research trips to Montserrat and includes plants that are not known to be ethnobotanically important to Montserratians. I designate my regional collections by a capital letter preceding the specimen number. The letter C refers to specimens collected in the Caribbean. The number in parentheses indicates how many specimens of a given taxon were collected at a site.

C-1 (3) *Cordia obliqua* Willd.; Clammy Cherry; Boraginaceae; Yacht Club Beach, 1 km south of Plymouth, coastal thicket; June 21, 1977.

C-2 (2) *Ipomoea pes-caprae* subsp. *brasiliensis* (L.) Ooststr.; Beach Morning Glory; Convolvulaceae; Yacht Club Beach, sea level; June 21, 1977.

C-3 (3) *Momordica charantia* L.; Pom Cooly; Cucurbitaceae; Yacht Club Beach, 1 km south of Plymouth, on coastal thicket border; June 21, 1977.

C-4 (3) *Acacia macracantha* Humb. & Bonpl.; Casha; Fabaceae; Yacht Club Beach, 1 km south of Plymouth, coastal thicket; June 21, 1977.

C-5 (3) *Antigonon leptopus* Hook. & Arn.; Coralita; Polygonaceae; Yacht Club Beach, 1 km south of Plymouth, coastal thicket; June 21, 1977.

C-6 (3) *Thunbergia alata* Boj. ex Sims; Golden Bells; Acanthaceae; Galway Road, 1 km west of Galway Soufriere; June 21, 1977.

C-7 (2) *Chenopodium ambrosioides* L.; Wormwood, Wormseed Weed; Chenopodi-

## Appendix

C-8 (6) *Trichachne insularis* (L.) Nees; Long Grass; Poaceae; Roadside, 1 km west of Galway Soufriere; June 21, 1977.

C-9 (3) *Abutilon indicum* (L.) Sweet; Rope Bush; Malvaceae; Thicket near old sugar mill 1 km west of Galway Soufriere; June 21, 1977.

C-10 (6) *Cajanus cajan* (L.) Huth; Pigeon Pea; Fabaceae; Roadside garden by old church 1 km west of Galway Soufriere; June 21, 1977.

C-11 (2) *Peperomia pellucida* (L.) Kth.; Inflammation Bush; Piperaceae; Near old church 1 km west of Galway Soufriere; June 21, 1977.

C-12 (3) *Solanum ficifolium* Ort.; Pigeonberry; Solanaceae; Near old church 1 km west of Galway Soufriere; June 21, 1977.

C-13 (4) *Asclepias curassavica* L.; Wild Ipecacuanha; Asclepiadaceae; Near old church 1 km west of Galway Soufriere; June 21, 1977.

C-14 (3) *Mammea americana* L.; Mammy Apple; Clusiaceae; 500 m from Galway Soufriere, roadside; June 21, 1977.

C-15 (4) *Amaranthus dubius* C. Mart. ex Thell.; aceae; Garden, 1 km west of Galway Soufriere; June 21, 1977. Spinach; Amaranthaceae; Galway Soufriere area, upper level garden; June 21, 1977.

C-16 (2) *Artemisia dracunculoides* Pursh.; Bush Tarragon; Asteraceae; Galway Soufriere area; June 21, 1977.

C-17 (3) *Stachytarpheta cayennensis* (L. C. Rich.) Vahl; Whorewine; Verbenaceae; Galway Soufriere area; June 21, 1977.

C-18 (3) *Piper amalago* L.; Malimbé; Piperaceae; Galway Soufriere area; June 21, 1977.

C-19 (3) *Clusia rosea* Jacq.; Wild Mammy Apple; Clusiaceae; Galway Soufriere area; June 21, 1977.

C-20 (2) *Palicourea crocea* (Sw.) Roem. & Schult.; Bois Puce; Rubiaceae; Galway Soufriere area; June 21, 1977.

C-21 (3) *Charianthus purpureus* D. Don; Wassard; Melastomaceae; Galway Soufriere area; June 21, 1977.

C-22 (6) *Eryngium foetidum* L.; Ramgoat Bush; Apiaceae; Galway Soufriere area; June 21, 1977.

C-23 (6) *Heteropogon contortus* (L.) Beauv. ex Roem. & Schult.; Goat Grass; Poaceae; Galway Soufriere area; June 21, 1977.

C-24 (1) *Torulinium odoratum* (L.) Hooper;

## Appendix

C-25 (4) *Cissus sicyoides* L.; Poison Wythe; Vitaceae; Galway Soufriere area; June 21, 1977.

Sedge; Cyperaceae; Galway Soufriere area; June 21, 1977.

C-26 (2) *Hippobroma longiflora* (L.) G. Don; Star of Bethlehem; Campanulaceae; Galway Soufriere area; June 21, 1977.

C-27 (3) *Theobroma cacao* L.; Cacao; Sterculiaceae; Galway Soufriere area, upper level garden; June 21, 1977.

C-28 (4) *Andropogon bicornis* L.; Sheep Grass; Poaceae; Galway Soufriere area; June 21, 1977.

C-29 (2) *Indigofera suffruticosa* Mill.; Indigo, Indigo Weed; Fabaceae; Galway Soufriere area; June 21, 1977.

C-30 (3) *Nepeta cataria* L.; Catnip, Catmint; Lamiaceae; Galway Soufriere area, upper level garden; June 21, 1977.

C-31 (3) *Cissus sicyoides* L.; Poison Wythe; Vitaceae; Woodlands, mesic hillside; June 21, 1977.

C-32 (3) *Arachis hypogaea* L.; Peanut, Groundnut; Fabaceae; R. Marland's garden, Woodlands; June 21, 1977.

C-33 (3) *Bernardia dichotoma* (Willd.) Muell.; Strongback; Euphorbiaceae; Woodlands, mesic rocky ravine; June 21, 1977.

C-34 (3) *Catharanthus roseus* (L.) G. Don; Twelve O'clock, Everyday Flower; Apocynaceae; Woodlands, mesic roadside ditch; June 21, 1977.

C-35 (3) *Merremia dissecta* (Jacq.) Hallier; Nora Bush; Convolvulaceae; Woodlands, mesic wooded hillside; June 21, 1977.

C-36 (2) *Matelea maritima* (Jacq.) Woodson; Man Tree; Asclepiadaceae; White River Ghaut, mesic rocky valley; June 22, 1977.

C-37 (3) *Croton flavens* L.; Rock Balsam; Euphorbiaceae; White River Ghaut, mesic rocky valley; June 22, 1977.

C-38 (3) *Abrus precatorius* L.; Jumbie Bead; Fabaceae; White River Ghaut, mesic rocky valley; June 22, 1977.

C-39 (2) *Bursera simaruba* (L.) Sarg.; Gum Bark; Burseraceae; White River Ghaut, mesic rocky valley; June 22, 1977.

C-40 (3) *Chiococca alba* (L.) Hitchc.; David's Root; Rubiaceae; White River Ghaut, mesic rocky valley; June 22, 1977.

C-41 (3) *Bursera simaruba* (L.) Sarg.; Gum Tree, Gum Bark; Burseraceae; White River Ghaut, mesic rocky valley; June 22, 1977.

## Appendix

C-42 (3) *Annona squamosa* L.; Sugar Apple; Annonaceae; White River Ghaut, mesic rocky valley; June 22, 1977.

C-43 (3) *Cordia curassavica* (Jacq.) Roem. & Schult.; Saize Bush; Boraginaceae; White River Ghaut, mesic rocky valley; June 22, 1977.

C-44 (2) *Sida ciliaris* L.; Twelve O'clock; Malvaceae; White River Ghaut, mesic rocky valley; June 22, 1977.

C-45 (3) *Tamarindus indica* L.; Tamarind; Fabaceae; White River Ghaut, mesic rocky valley; June 22, 1977.

C-46 (6) *Jatropha curcas* L.; Body Cutter; Euphorbiaceae; White River Ghaut, mesic rocky valley; June 22, 1977.

C-47 (3) *Melia azedarach* L.; Lilac; Meliaceae; White River Ghaut, mesic rocky valley; June 22, 1977.

C-48 (2) *Anacardium occidentale* L.; Cashew Tree; Anacardiaceae; White River Ghaut, mesic rocky valley; June 22, 1977.

C-49 (3) *Jatropha gossypifolia* L.; French Body Cutter; Euphorbiaceae; White River Ghaut, mesic rocky valley; June 22, 1977.

C-50 (3) *Pithecellobium saman* (Jacq.) Benth.; Rainfall Tree; Fabaceae; White River Ghaut, mesic rocky valley; June 22, 1977.

C-51 (2) *Datura innoxia* Mill.; Painkiller; Solanaceae; White River Valley, field border; June 22, 1977.

C-52 (3) *Crescentia cujete* L.; Calabash; Bignoniaceae; Salem Village, dooryard garden; June 23, 1977.

C-53 (3) *Cassia obtusifolia* L.; Money Bush; Fabaceae; White River near Great Alp Falls, mesic woodland; June 23, 1977.

C-54 (3) *Adelia ricinella* L.; Batroot; Euphorbiaceae; White River Valley; June 23, 1977.

C-55 (2) *Piscidia piscipula* (L.) Sarg.; Dogwood; Fabaceae; White River near Great Alp Falls, mesic woodland; June 23, 1977.

C-56 (2) *Passiflora foetida* L.; Baby Honeysuckle; Passifloraceae; White River Ghaut near Great Alp Falls, mesic woodland; June 23, 1977.

C-57 (5) *Trichachne insularis* (L.) Nees; Long Grass; Poaceae; White River Ghaut, grassy slope; June 23, 1977.

C-58 (2) *Mimosa pudica* L.; Shame Bush; Fabaceae; Great Alp Falls area, mesic slope; June 23, 1977.

C-59 (2) *Cedrela mexicana* M. J. Roem.; Red

Cedar; Meliaceae; Great Alp Falls area, mesic woodland; June 23, 1977.

C-60 (2) *Origanum vulgare* L.; Oregano, Balsam; Lamiaceae; Public market, Plymouth; June 24, 1977.

C-61 (2) *Nectandra membranacea* (Sw.) Griseb.; Sweetwood, Southernwood; Lauraceae; Public market, Plymouth; June 24, 1977.

C-62 (2) *Rosmarinus officinalis* L.; Rosemary; Lamiaceae; Public market, Plymouth; June 24, 1977.

C-63 (2) *Cymbopogon citratus* (DC. ex Nees) Stapf; Fever Grass, Lemon Grass; Poaceae; Public market, Plymouth; June 24, 1977.

C-64 (2) *Origanum vulgare* L.; Oregano; Lamiaceae; Public market, Plymouth; June 24, 1977.

C-65 (2) *Nepeta cataria* L.; Catmint, Catnip; Lamiaceae; Public market, Plymouth; June 24, 1977.

C-66 (2) *Spondias purpurea* L.; West Indian Plum; Anacardiaceae; Public market, Plymouth; June 24, 1977.

C-67 (2) *Arachis hypogaea* L.; Groundnut, Peanut, Monkey nut; Fabaceae; Public market, Plymouth; June 24, 1977.

C-68 (2) *Origanum majorana* L.; Sweet Marjoram; Lamiaceae; Public market, Plymouth; June 24, 1977.

C-69 (2) *Mentha piperita* L.; Peppermint; Lamiaceae; Public market, Plymouth; June 24, 1977.

C-70 (2) *Tanacetum vulgare* L.; Tansy; Asteraceae; Tar River Estate, dooryard garden; June 24, 1977.

C-71 (1) *Portulaca oleracea* L.; Pursly, Purslane; Portulacaceae; Tar River Estate, dooryard garden; June 24, 1977.

C-72 (3) *Malpighia punicifolia* L.; Barbados Cherry; Malpighiaceae; Tar River Estate, dooryard garden; June 24, 1977.

C-73 (3) *Syzygium jambos* (L.) Alston; Rose Plum; Myrtaceae; Tar River Estate, dooryard garden; June 24, 1977.

C-74 (3) *Cymbopogon citratus* (DC. ex Nees) Stapf; Lemon Grass; Poaceae; Tar River Estate, roadside; June 24, 1977.

C-75 (2) *Psidium guajava* L.; Guava; Myrtaceae; Tar River Estate, dooryard garden; June 24, 1977.

C-76 (1) *Merremia umbellata* (L.) H. Hallier; Hog Vine; Convolvulaceae; Woodlands, dooryard garden; June 24, 1977.

C-77 (3) *Chiococca alba* (L.) Hitchc.; David's

## Appendix

Root; Rubiaceae; Centre Hills, mesic hillside; June 27, 1977.

C-78 (2) *Citrus reticulata* Blanco; Tangerine; Rutaceae; Centre Hills, mesic hillside, upper level garden; June 27, 1977.

C-79 (7) *Polypodium heterophyllum* L.; Climbing Tongue Fern; Polypodiaceae; Centre Hills, mesic forest, on *Mangifera indica* L.; June 27, 1977.

C-80 (2) *Citrus aurantium* L.; Swivel Sweet; Rutaceae; Centre Hills, upper level garden; June 27, 1977.

C-81 (2) *Castilla elastica* Sessé; Wild Rubber Tree; Moraceae; Centre Hills, upper level garden; June 27, 1977.

C-82 (4) *Piper dilatatum* Rich.; Elder Bush, Rock Bush; Piperaceae; Centre Hills, mesic forest; June 27, 1977.

C-83 (3) *Ricinus communis* L.; White Castor Nut; Euphorbiaceae; Woodlands, dooryard garden; June 27, 1977.

C-84 (5) *Zanthoxylum monophyllum* (Lam.) P. Wils.; Yellow Prickle; Rutaceae; Centre Hills, mesic forest; June 27, 1977.

C-85 (3) *Mammea americana* L.; Mammy Apple; Clusiaceae; Centre Hills, upper level garden; June 27, 1977.

C-86 (2) *Helianthus hirsutus* Raf.; Sage; Asteraceae; Centre Hills, upper level garden; June 27, 1977.

C-87 (1) *Bursera simaruba* (L.) Sarg.; Gum tree; Burseraceae; Centre Hills, mesic forest; June 27, 1977.

C-88 (3) *Lantana camara* var. *aculeata* (L.) Moldenke; Red Sage; Verbenaceae; Centre Hills, mesic forest; June 27, 1977.

C-89 (7) *Manihot esculenta* Crantz; Cassava; Euphorbiaceae; Salem Village, dooryard garden; June 27, 1977.

C-90 (5) *Peperomia pellucida* (L.) Kth.; Inflammation Bush; Piperaceae; Centre Hills, mesic forest edge; June 27, 1977.

C-91 (10) *Phoradendron trinervium* (Lam.) Griseb.; No Mammy; Loranthaceae; Centre Hills, mesic forest, growing on *Bursera simaruba* (L.) Sarg.; June 27, 1977.

C-92 (3) *Manihot esculenta* Crantz; Cassava; Euphorbiaceae; Centre Hills, dooryard garden; June 27, 1977.

C-93 (4) *Cedrela mexicana* M. J. Roem.; Red Cedar; Meliaceae; Olveston, McChesney Estate, mesic hillside; June 27, 1977.

C-94 (1) *Cordia curassavica* (Jacq.) Roem. & Schult.; Cow's Tongue; Boraginaceae; Olveston, thorn forest; June 27, 1977.

C-95 (3) *Vetiveria zizanioides* (L.) Nash; Sweet Grass; Poaceae; Olveston, roadside ditch; June 27, 1977.

C-96 (1) *Pithecellobium saman* (Jacq.) Benth.; Raintree; Fabaceae; Olveston, fence row; June 27, 1977.

C-97 (3) *Commelina diffusa* Burm.; White Frenchweed; Commelinaceae; Olveston, mesic thicket; June 27, 1977.

C-98 (2) *Chenopodium ambrosioides* L.; Worm Oil Tree; Chenopodiaceae; Salem, dooryard garden; June 28, 1977.

C-99 (3) *Croton flavens* L.; Sweet Balsam; Euphorbiaceae; White River Valley; June 29, 1977.

C-100 (3) *Triphasia trifolia* (Burm. f.) P. Wils.; Sweet Lime; Rutaceae; White River Valley, mesic stream border; June 29, 1977.

C-101 (3) *Portulaca oleracea* L.; Pusly; Portulacaceae; White River Valley, field; June 29, 1977.

C-102 (3) *Mimosa pudica* L.; Strongback; Fabaceae; White River Valley; June 29, 1977.

C-103 (1) *Hymenaea courbaril* L.; Locust Tree; Fabaceae; White River Valley, by stream; June 29, 1977.

C-104 (3) *Cordia nitida* Vahl; Leley, Red Manjack; Boraginaceae; White River Valley, xeric hillside; June 29, 1977.

C-105 (4) *Cassia bicapsularis* L.; Money Money Tree; Fabaceae; White River Valley, mesic hillside; June 29, 1977.

C-106 (4) *Eucalyptus resinifera* J. E. Smith; Eucalyptus; Myrtaceae; Near Emerald Isle Hotel, roadside; June 29, 1977.

C-107 (4) *Oncidium altissimum* (Jacq.) Sw.; Butterfly Flower; Orchidaceae; White River Valley, on ground, mesic hill; some on trees and rocks; June 29, 1977.

C-108 (3) *Passiflora foetida* L.; Baby Honeysuckle; Passifloraceae; White River Valley, hillside thicket; June 29, 1977.

C-109 (1) *Annona muricata* L.; Soursop; Annonaceae; Plymouth, dooryard garden; June 29, 1977.

C-110 (3) *Tragia volubilis* L.; Cow Itch, Vine Nettle; Euphorbiaceae; Woodlands, mesic hillside; July 2, 1977.

C-111 (3) *Capsicum annuum* L.; Hotpepper Bush; Solanaceae; Woodlands, dooryard garden; July 2, 1977.

C-112 (3) *Melia azedarach* L.; Lilac; Meliaceae; Woodlands, mesic hillside forest; July 2, 1977.

C-113 (3) *Adelia ricinella* L.; Batroot; Euphor-

## Appendix

biaceae; Woodlands, mesic hillside; July 2, 1977.
C-114 (7) *Melicoccus bijugatus* Jacq.; Genip; Sapindaceae; Woodlands, mesic hillside; July 2, 1977.
C-115 (3) *Licaria triandra* (Sw.) Kosterm.; Sweetwood; Lauraceae; Woodlands, mesic forest; July 2, 1977.
C-116 (1) *Melia azedarach* L.; Redwood; Meliaceae; Woodlands, mesic forest; July 2, 1977.
C-117 (6) *Mangifera indica* L.; Mango; Anacardiaceae; Woodlands, dooryard garden; July 2, 1977.
C-118 (5) *Colocasia esculenta* (L.) Schott; Eddee; Araceae; Woodlands, dooryard garden; July 2, 1977.
C-119 (3) *Colocasia esculenta* (L.) Schott; Dasheen; Araceae; Woodlands, dooryard garden; July 2, 1977.
C-120 (3) *Digitalis purpurea* L.; Heart Bush; Scrophulariaceae; Woodlands; July 2, 1977.
C-121 *Centrosema virginianum* L. Benth.; Wild Pea; Fabaceae; Woodlands, dooryard garden; July 2, 1977.
C-122 (1) *Colocasia esculenta* (L.) Schott; Dasheen; Araceae; Woodlands, dooryard garden; July 2, 1977.
C-123 (3) *Hibiscus sabdariffa* L.; Sorrel; Malvaceae; Woodlands, mesic hillside; July 2, 1977.
C-124 (5) *Trichachne insularis* (L.) Nees; Corn Grass; Poaceae; Woodlands, mesic hillside; July 2, 1977.
C-125 (2) *Myrcia splendens* (Sw.) DC.; Red Rodwood; Myrtaceae; Woodlands, mesic woods; July 2, 1977.
C-126 (3) *Euphorbia heterophylla* L.; Jacob's Ladder; Euphorbiaceae; Woodlands, mesic hillside; July 2, 1977.
C-127 (3) *Emilia sonchifolia* (L.) DC.; Rabbit Food; Asteraceae; Woodlands, mesic hillside; July 2, 1977.
C-128 (3) *Sida ciliaris* L.; Twelve O'clock; Malvaceae; Woodlands, mesic hillside; July 2, 1977.
C-129 (3) *Saccharum officinarum* L.; Sugarcane; Poaceae; Woodlands, dooryard garden; July 2, 1977.
C-130 (5) *Artocarpus altilis* (Parkinson) Fosb.; Breadfruit; Moraceae; Woodlands, mesic hillside; July 2, 1977.
C-131 (4) *Diospyros revoluta* Poir.; Black Apple; Ebenaceae; Woodlands, mesic hillside; July 2, 1977.
C-132 (3) *Inga laurina* (Sw.) Willd.; Spanish Oak;

# Appendix

C-133 (2) *Euterpe globosa* Gaertn.; Cabbage Palm; Arecaceae; Woodlands, mesic hillside; Fabaceae; Woodlands, mesic hillside; July 2, 1977.

C-133 (2) *Euterpe globosa* Gaertn.; Cabbage Palm; Arecaceae; Woodlands, mesic hillside; July 2, 1977.

C-134 (2) *Solanum nigrum* L.; Ink Balls; Solanaceae; Woodlands, mesic wooded hillside; July 2, 1977.

C-135 (2) *Solanum nodiflorum* Dunal sensu lato; Agouman; Solanaceae; Woodlands, mesic wooded hillside; July 2, 1977.

C-136 (2) *Erigeron canadensis* L.; Wild Tarragon; Asteraceae; Woodlands, dooryard garden; July 2, 1977.

C-137 (5) *Leonotis nepetifolia* (L.) Ait.; Lord Lovington; Lamiaceae; Woodlands, R. Marland's pasture; July 2, 1977.

C-138 (4) *Sambucus simpsonii* Rehd.; Elder; Caprifoliaceae; Woodlands, mesic hillside; July 2, 1977.

C-139 (1) *Stachytarpheta cayennensis* (L. C. Rich.) Vahl; Devil's Tail; Verbenaceae; Woodlands, mesic hillside; July 2, 1977.

C-140 (2) *Ruellia tuberosa* L.; Double Bit; Acanthaceae; Woodlands, mesic hillside; July 2, 1977.

C-141 (4) *Achyranthes indica* (L.) Mill.; Devil's Horse Whip; Amaranthaceae; Woodlands, mesic hillside; July 2, 1977.

C-142 (3) *Cassia occidentalis* L.; Stinking Bush; Fabaceae; Woodlands, mesic hillside; July 2, 1977.

C-143 (3) *Melicoccus bijugatus* Jacq.; Genip; Sapindaceae; Woodlands, mesic hillside; July 2, 1977.

C-144 (1) *Sechium edule* (Jacq.) L.; Christophine; Cucurbitaceae; Woodlands, dooryard garden; July 2, 1977.

C-145 (2) *Leonotis nepetifolia* (L.) Ait.; Lord Lovington; Lamiaceae; Woodlands, pasture; July 2, 1977.

C-146 (3) *Bryophyllum pinnatum* (Lam.) Oken; Love Bush; Crassulaceae; Woodlands, R. Marland's pasture; July 2, 1977.

C-147 (4) *Crotalaria quinquefolia* L.; Pop Bush; Fabaceae; Seaside at Plymouth; July 2, 1977.

C-148 (1) *Crotalaria incana* L.; Rattle Bush, Shake Shake; Fabaceae; Seaside at Plymouth; July 2, 1977.

C-149 (3) *Passiflora edulis* Sims; Passionfruit; Passifloraceae; Woodlands, dooryard garden of Alic Inglis; July 2, 1977.

C-150 (2) *Hippomane mancinella* L.; Manchineel;

## Appendix

Euphorbiaceae; Beach by Vue Pointe Hotel; July 6, 1977.
C-151 (1) *Cassia fistula* L.; Golden Shower; Fabaceae; Woodlands, upper level garden border; July 6, 1977.
C-152 (3) *Plumeria rubra* L.; Red Frangipani; Apocynaceae; Woodlands, R. Marland's yard; July 6, 1977.
C-153 (3) *Plumeria alba* L.; Wild Frangipani; Apocynaceae; Woodlands, R. Marland's yard; July 6, 1977.
C-154 (3) *Euphorbia tirucalli* L.; Pencil Plant; Euphorbiaceae; Lime Kiln Beach, hedge around house and dooryard garden; July 6, 1977.
C-155 (3) *Sansevieria metallica* Gerome & Labroy; Coretor; Liliaceae; Lime Kiln Beach, dooryard garden; July 6, 1977.
C-156 (1) *Clusia rosea* Jacq.; Wild Mammy Apple; Clusiaceae; Wooded hill above Salem; July 6, 1977.
C-157 (1) *Cassia planisiliqua* L.; Wild Pea; Fabaceae; Clearing above Salem; July 6, 1977.
C-158 (3) *Thevetia peruviana* (Pers.) K. Schum.; Luckynut, Luckyseed; Apocynaceae; Salem, dooryard garden; July 6, 1977.
C-159 (3) *Acacia tortuosa* (L.) Willd.; Friendego; Fabaceae; Wooded hill above Salem; July 6, 1977.
C-160 *Trichachne insularis* (L.) Nees; Long Grass; Poaceae; Salem, garden border; July 6, 1977.
C-161 (3) *Macfadyena unguis-cati* (L.) A. Gentry; Cat's Claw; Bignoniaceae; Mesic woods above Salem; July 6, 1977.
C-162 (3) *Cordia alliodora* (Ruiz & Pav.) Oken; Seepwood; Boraginaceae; Mesic woods above Salem; July 6, 1977.
C-163 (3) *Byrsonima spicata* (Cav.) DC.; Shoemaker Bark; Malpighiaceae; Mesic woods above Salem; July 6, 1977.
C-164 *Lantana reticulata* Pers.; Measle Bush; Verbenaceae; Mesic hill above Salem; July 6, 1977.
C-165 *Ervatamia cumingiana* (A. DC.) Markgr.; Milky; Apocynaceae; Mesic hill above Salem; July 6, 1977.
C-166 (3) *Sida ciliaris* L.; Twelve O'clock; Malvaceae; Woods above Salem; July 6, 1977.
C-167 (5) *Alpinia zerumbet* (Persoon) Burtt & R. R. Smith; Shell Plant; Zingiberaceae; Dooryard garden, Salem; July 6, 1977.
C-168 (3) *Simarouba amara* Aubl.; Tom Tah; Simaroubaceae; Mesic forest above Salem; July 6, 1977.

C-169 (4) *Phyllanthus anderssonii* Muell.; Red Iron Bark; Euphorbiaceae; Mesic hillside above Salem; July 6, 1977.

C-170 (3) *Cassia bicapsularis* L.; Money Plant; Fabaceae; Mesic hillside above Salem; July 6, 1977.

C-171 (3) *Randia aculeata* L.; Goat Hahn; Rubiaceae; Mesic hillside above Salem; July 6, 1977.

C-172 (4) *Nectandra coriacea* (Sw.) Griseb.; Sweetwood; Lauraceae; Mesic hillside above Salem; July 6, 1977.

C-173 (3) *Simarouba amara* Aubl.; Snakewood; Simaroubaceae; Mesic hillside above Salem; July 6, 1977.

C-174 *Tournefortia filiflora* Griseb.; Elder; Boraginaceae; Mesic hillside above Salem; July 6, 1977.

C-175 (3) *Annona montana* Macf.; Wild Soursop; Annonaceae; Mesic hillside above Salem; July 6, 1977.

C-176 (4) *Ceiba pentandra* (L.) Gaertn.; Silkcotton; Bombacaceae; Mesic hillside above Salem; July 6, 1977.

C-177 (3) *Gliricidia sepium* (Jacq.) Steud.; Gorey Cedar; Fabaceae; Mesic hillside above Salem; July 6, 1977.

C-178 (1) *Origanum vulgare* L.; Oregano; Lamiaceae; Salem, mesic hillside; July 6, 1977.

C-179 (3) *Fleurya aestuans* (L.) Miq.; Nettle; Urticaceae; Mesic hillside above Salem; July 6, 1977.

C-180 (4) *Lycopodium clavatum* L.; Snakeweed; Lycopodiaceae; Mesic hillside on Cavalla Hill, 457 m; July 6, 1977.

C-181 (3) *Stachytarpheta cayennensis* (L. C. Rich.) Vahl; Whorewine; Verbenaceae; Mesic hillside; Paradise; July 11, 1977.

C-182 (4) *Wedelia trilobata* (L.) Hitchc.; Pasture Sage; Asteraceae; Mesic pasture at Paradise; July 11, 1977.

C-183 (1) *Bidens pilosa* L.; Spanish Needles, Duppy Needles; Asteraceae; Paradise, mesic roadside; July 11, 1977.

C-184 (1) *Eleusine indica* (L.) Gaertn.; Dutchgrass; Poaceae; Old roadbed at Paradise; July 11, 1977.

C-185 (3) *Cynodon dactylon* (L.) Pers.; Hardgrass; Poaceae; Old roadbed at Paradise; July 11, 1977.

C-186 (1) *Galactia filiformis* Benth.; Strongback; Fabaceae; Roadside at Paradise; July 11, 1977.

C-187 (3) *Rubus rosifolius* J. E. Smith; Raspberry; Rosaceae; Hillside thicket at Paradise; July 11, 1977.

## Appendix

C-188 (3) *Canna edulis* Ker-Gawl.; Mountain Porridge, Toiseloisemoise, Toloma Food; Cannaceae; Mesic hillside at Paradise; July 11, 1977.

C-189 (2) *Eupatorium villosum* Sw.; Sage; Asteraceae; Mesic hillside at Paradise; July 11, 1977.

C-190 (2) *Allium fistulosum* L.; Welsh Onion; Liliaceae; Mesic hillside at Paradise; July 11, 1977.

C-191 (2) *Abutilon hirtum* (Lam.) Sweet; Burry Bark; Malvaceae; Mesic hillside at Paradise; July 11, 1977.

C-192 (3) *Turbina corymbosa* (L.) Raf.; Christmas Bush; Convolvulaceae; Mesic hillside at Paradise; July 11, 1977.

C-193 (4) *Cissus sicyoides* L.; Skip Rope; Vitaceae; Mesic hillside at Paradise; July 11, 1977.

C-194 (4) *Flacourtia jangomas* (Lour.) Raeusch.; Governor's Plum, Java Plum; Flacourtiaceae; Mesic hillside at Paradise; July 11, 1977.

C-195 (3) *Monstera adansonii* Schott; Sarsaparilla; Araceae; Paradise, mesic fence row; July 11, 1977.

C-196 (4) *Datura innoxia* Mill.; Wonga; Solanaceae; Trants Estate, old field; July 11, 1977.

C-197 (2) *Annona squamosa* L.; Sweetsop; Annonaceae; Trants Estate, dooryard garden; July 11, 1977.

C-198 (4) *Tabebuia pallida* (Lindl.) Miers; White Cedar; Bignoniaceae; Trants Estate, fence row; July 11, 1977.

C-199 (1) *Hylocereus trigonus* (Haw.) Safford; Wild Strawberry; Cactaceae; Woodlands, old estate; July 11, 1977.

C-200 (3) *Swietenia macrophylla* King; Honduras Mahogany; Meliaceae; Woodlands, mesic hillside; July 11, 1977.

C-201 (4) *Euphororbia maculata* L. var. *thymifolia* L.; Eyebright; Euphorbiaceae; Trants Estate, mesic hillside; July 18, 1977.

C-202 (4) *Piscidia piscipula* (L.) Sarg.; Dogwood; Fabaceae; Harris Village, mesic hillside; July 18, 1977.

C-203 (3) *Blighia sapida* König; Akee; Sapindaceae; Dooryard garden below Bugby Hole; July 18, 1977.

C-204 (3) *Croton flavens* L.; Rock Balsam; Euphorbiaceae; Bugby Hole; July 18, 1977.

C-205 (6) *Plumeria bahamensis* Urban; Wild Frangipani; Apocynaceae; Bugby Hole, xeric scrub forest; July 18, 1977.

C-206 (3) *Pogonia rosea* (Lindl.) Hemsl.; Orchidaceae; Chance Peak, rainforest, epiphytic on tree, 853 m; July 19, 1977.

C-207 (1) *Anthurium willdenowii* Kth.; Araceae; Chance Peak, rainforest, epiphytic on tree, 853 m; July 19, 1977.

C-208 (3) *Eleocharis mutata* (L.) Roem. & Schult.; Cyperaceae; Chance Pond, marsh, 914 m; July 19, 1977.

C-209 (2) *Anthurium cordatum* (L.) Schott; Araceae; Chance Peak, 884 m; July 19, 1977.

C-210 (2) *Anthurium lanceolatum* Kth.; Araceae; Chance Peak, rainforest, 853 m; July 19, 1977.

C-211 (5) *Cephaelis swartzii* DC.; Bois Marguerite, Ipéca Bâtard; Rubiaceae; Chance Peak, rainforest, 823 m; July 19, 1977.

C-212 (4) *Heliconia caribaea* Lam.; Balisier; Musaceae; Chance Peak, rainforest, 823 m; July 19, 1977.

C-213 (1) *Monstera adansonii* Schott; Seguine Couleuvre, Liane Percée; Araceae; Chance Peak, rainforest, 610 m; July 19, 1977.

C-214 (3) *Campyloneurum phyllitidis* (L.) C. Presl; Polypodiaceae; Chance Peak, rainforest, 610 m; July 19, 1977.

C-215 (2) *Vittaria lineata* (L.) Smith; Polypodiaceae; Chance Peak, rainforest, 610 m; July 19, 1977.

C-216 (2) *Selaginella substipitata* Spring; Selaginellaceae; Chance Peak, rainforest, 610 m; July 19, 1977.

C-217 (2) *Selaginella eatoni* Hieron; Selaginellaceae; Chance Peak, rainforest, 610 m; July 19, 1977.

C-218 (4) *Voyria tenella* Guild; Muguet Bleu; Gentianaceae; Chance Peak, rainforest, 610 m; July 19, 1977.

C-219 (5) *Psilotum nudum* (L.) Griseb.; Tickleweed; Psilotaceae; On *Mangifera indica* L.; Chance Peak, edge of field, 305 m; July 19, 1977.

C-220 (4) *Trichachne insularis* (L.) Nees; Corn Grass; Poaceae; Woodlands, roadside ditch; July 21, 1977.

C-221 (3) *Codiaeum variegatum* (L.) Adr. Juss.; Euphorbiaceae; Woodlands, dooryard garden; July 21, 1977.

C-222 (2) *Anthurium willdenowii* Kth.; Araceae; Centre Hills, rainforest; July 21, 1977.

C-223 (3) *Castilla elastica* Sessé; Wild Rubber Tree; Moraceae; Centre Hills, rainforest; July 21, 1977.

C-224 (3) *Bambusa vulgaris* Schrad. ex Wendl.;

## Appendix

Bamboo; Poaceae; Centre Hills, rainforest; July 21, 1977.

C-225 (4) *Ficus aurea* Nutt.; Evil Tree; Moraceae; Woodlands, rainforest; July 21, 1977.

C-226 (5) *Zingiber officinale* Roscoe; Ginger; Zingiberaceae; Woodlands, upper level garden; July 21, 1977.

C-227 (3) *Heliconia subulata* Ruiz & Pav.; Bird of Paradise; Musaceae; Woodlands, R. Marland's yard; July 21, 1977.

C-228 (3) *Citrus aurantifolia* (Christm.) Swingle; Lime; Rutaceae; Woodlands, dooryard garden; July 21, 1977.

C-229 (3) *Nerium oleander* L.; Oleander; Apocynaceae; Woodlands, dooryard garden; July 21, 1977.

C-230 (1) *Acalypha virgata* (L.) Sw.; Ti Codinde; Euphorbiaceae; Woodlands, edge of mesic forest; July 21, 1977.

C-231 (1) *Hibiscus sabdariffa* L.; Sorrel, Roselle; Malvaceae; Salem dooryard garden of Alfred Payne; July 21, 1977.

C-232 (1) *Cecropia peltata* L.; Trumpeter, Trumpet Tree; Moraceae; Salem, dooryard garden of Alfred Payne; July 21, 1977.

C-233 (1) *Carissa macrocarpa* (Ecklon) A. DC.; Natal Plum; Apocynaceae; Salem, dooryard garden; July 21, 1977.

C-234 (3) *Amaranthus dubius* C. Mart. ex Thell.; Spinach; Amaranthaceae; Salem, dooryard garden of Alfred Payne; July 21, 1977.

C-235 (2) *Euphorbia oerstediana* Klotzsch & Garcke; Euphorbiaceae; Salem, roadside; July 21, 1977.

C-236 (3) *Abrus precatorius* L.; Jumbie Bead; Fabaceae; Salem, fence row; July 21, 1977.

C-237 (2) *Xanthosoma brasiliense* (Desf.) Engler; Crackers; Araceae; Great Alp Falls, mesic canyon; July 21, 1977.

C-238 (3) *Eugenia sessiliflora* Vahl; Myrtaceae; St. George's Hill, mesic slope; July 23, 1977.

C-239 (3) *Syzygium malaccense* (L.) Merr. & Perry; Rose Apple; Myrtaceae; St. George's Hill, dooryard garden; July 23, 1977.

C-240 (3) *Diospyros revoluta* Poir.; Black Apple; Ebenaceae; Salem, mesic hillside; July 23, 1977.

C-241 (3) *Punica granatum* L.; Pomegranate; Punicaceae; Olveston, McChesney Estate; July 23, 1977.

C-242 (3) *Moringa oleifera* Lam.; Horseradish Tree; Moringaceae; Olveston, McChesney Estate; July 23, 1977.

C-243 (3) *Bixa orellana* L.; Lipstick Plant; Bixaceae; Olveston, McChesney Estate; July 23, 1977.

C-244 (3) *Malpighia punicifolia* L.; Barbados Cherry, Acerola Cherry; Malpighiaceae; Olveston, dooryard garden; July 23, 1977.

C-245 (6) *Cinnamomum zeylanicum* Bl.; Cinnamon; Lauraceae; Olveston, McChesney Estate; July 23, 1977.

C-246 (3) *Macadamia ternifolia* F. Muell.; Macadamia; Proteaceae; Olveston, McChesney Estate; July 23, 1977.

C-247 (2) *Cassytha filiformis* L.; Love Vine; Lauraceae; Olveston, McChesney Estate; July 23, 1977.

C-248 (3) *Haematoxylon campechianum* L.; Logwood; Fabaceae; Olveston, McChesney Estate; July 23, 1977.

C-249 (3) *Artocarpus heterophyllus* Lam.; Jackfruit; Moraceae; Olveston, McChesney Estate; July 23, 1977.

C-250 (3) *Phyllanthus acidus* (L.) Skeels; Gooseberry Tree; Euphorbiaceae; Olveston, McChesney Estate; July 23, 1977.

C-251 (3) *Pithecellobium unguis-cati* (L.) Mart.; Bread and Cheese; Fabaceae; Olveston, McChesney Estate; July 23, 1977.

C-252 (3) *Spondias cytherea* Sonner; Golden Apple; Anacardiaceae; Olveston, McChesney Estate; July 23, 1977.

C-253 (3) *Spondias mombin* L.; Hog Plum; Anacardiaceae; Olveston, McChesney Estate; July 23, 1977.

C-254 (2) *Flacourtia jangomas* (Lour.) Raeusch.; Governor's Plum; Flacourtiaceae; Olveston, McChesney Estate; July 23, 1977.

C-255 (5) *Pterocarpus indicus* Willd.; Burma Rosewood; Fabaceae; Shoe Rock, by old well on abandoned estate; July 29, 1977.

C-256 (6) *Guaiacum officinale* L.; Lignum Vitae; Zygophyllaceae; Shoe Rock, by old well on abandoned estate; July 29, 1977.

C-257 (3) *Guaiacum officinale* L.; Lignum Vitae; Zygophyllaceae; Shoe Rock, by old well on abandoned estate; July 29, 1977.

C-258 (3) *Capparis flexuosa* (L.) L.; Rat Bean, Caper Tree; Capparaceae; Shoe Rock, by old well on abandoned estate; July 29, 1977.

C-259 (1) *Guettarda elliptica* Sw.; Rat Apple; Rubiaceae; Shoe Rock, by old well on abandoned estate; July 29, 1977.

C-260 (3) *Jatropha gossypifolia* L.; French Body Cutter; Euphorbiaceae; Shoe Rock, by

## Appendix

old well on abandoned estate; July 29, 1977.

C-261 (3) *Croton flavens* L.; Rock Balsam; Euphorbiaceae; Shoe Rock, by old well on abandoned estate; July 29, 1977.

C-262 (2) *Thymus vulgaris* L.; Thyme; Lamiaceae; Paradise, cultivated field; July 30, 1977.

C-263 (3) *Psidium guajava* L.; Guava; Myrtaceae; Paradise, mesic thicket; July 30, 1977.

C-264 (2) *Solanum erianthum* D. Don; Jumbie Berry; Solanaceae; Paradise, mesic hillside; July 30, 1977.

C-265 (2) *Canna generalis* Bailey; Toloma; Cannaceae; Paradise, mesic hillside; July 30, 1977.

C-266 (2) *Ficus nymphaeifolia* Miller; Black Fig, Banyan; Moraceae; Paradise, mesic forest; July 30, 1977.

C-267 (2) *Smilax cumanensis* Humb. & Bonpl. ex Willd.; Wild Sarsaparilla; Smilacaceae; Paradise, growing as a vine on *Mangifera indica*; July 30, 1977.

C-268 (4) *Mimusops coriacea* (A. DC.) Miq.; Aphrodite's Apple, Passion Apple, Yellow Sapote; Sapotaceae; Paradise, mesic forest; July 30, 1977.

C-269 (3) *Morinda citrifolia* L.; Painkiller; Rubiaceae; Blackburn Airport, roadside; July 30, 1977.

C-270 (4) *Chiococca alba* (L.) Hitchc.; David's Root; Rubiaceae; Great Alp Falls, vine in mesic forest; July 31, 1977.

C-271 (2) *Miconia prasina* (Sw.) DC.; Hogwood; Melastomaceae; Galway Soufriere area, mesic hillside; July 31, 1977.

C-272 (4) *Delonix regia* (Boj. ex Hook.) Raf.; Flamboyant; Fabaceae; Galway Soufriere area, near old sugar mill; July 31, 1977.

C-273 (1) *Pitcairnia angustifolia* Ait.; Bromeliaceae; Galway Soufriere area in dry rocky soil near soufriere; July 31, 1977.

C-274 (3) *Gossypium barbadense* L.; Sea Island Cotton; Malvaceae; White River Valley, cultivated field; July 31, 1977.

C-275 (2) *Malvastrum americanum* (L.) Torr.; Mauve Savane; Malvaceae; White River Valley, edge of mesic field; July 31, 1977.

C-276 (3) *Hibiscus vitifolius* L.; Rope Plant; Malvaceae; White River Valley, edge of mesic field; July 31, 1977.

C-277 (3) *Cassia glandulosa* var. *swartzii* (Wikstr.) J. F. Macbr.; Wild Tamarind; Fabaceae; White River Valley, edge of river; July 31, 1977.

C-278 (3) *Lippia nodiflora* (L.) Michx.; Man Better Man; Verbenaceae; White River Valley, edge of mesic field; July 31, 1977.

C-279 (3) *Aloe vera* (L.) Burm. f.; Sintibibi, Bitter Aloe, Aloe; Liliaceae; White River Valley; July 31, 1977.

C-280 (4) *Coccoloba uvifera* (L.) L.; Seagrape; Polygonaceae; White River Valley, near Radio Antilles building; July 31, 1977.

C-281 (3) *Mammea americana* L.; Mammy; Clusiaceae; Woodlands, mesic hillside forest; August 1, 1977.

C-282 (3) *Inga laurina* (Sw.) Willd.; Spanish Oak; Fabaceae; Woodlands, mesic hillside forest; August 1, 1977.

C-283 (1) *Guatteria caribaea* Urban; Corossol Marron, Bois Violin; Annonaceae; Woodlands, mesic hillside forest; August 1, 1977.

C-284 (2) *Cassia bicapsularis* L.; Money Money Bush; Fabaceae; Salem, dooryard garden of Alfred Payne; August 1, 1977.

C-285 (1) *Cassia occidentalis* L.; Stinking Bush; Fabaceae; Salem, dooryard garden of Alfred Payne; August 1, 1977.

C-286 (1) *Crotalaria lotifolia* L.; Coot Weed; Fabaceae; Salem, dooryard garden of Alfred Payne; August 1, 1977.

C-287 (4) *Stachytarpheta jamaicensis* (L.) Vahl; Eyebright; Verbenaceae; Salem, weedy mesic hillside; August 1, 1977.

C-288 (4) *Beslaria lutea* L.; Bois Gleau, Honey Bush; Gesneriaceae; Mesic hill above Salem; August 1, 1977.

C-289 (3) *Chiococca alba* (L.) Hitchc.; David's Root; Rubiaceae; Mesic forest above Salem; August 1, 1977.

C-290 (2) *Theobroma cacao* L.; Cocoa Tree; Sterculiaceae; Upper level garden above Salem; August 1, 1977.

C-291 (2) *Psidium guajava* L.; Guava; Myrtaceae; Mesic hillside above Salem; August 1, 1977.

C-292 (1) *Cephaelis axillaris* Sw.; Ipéca Bâtard, Bois Marguerite; Rubiaceae; Mesic hillside above Salem; August 1, 1977.

C-293 (1) *Mangifera indica* L.; Mango; Anacardiaceae; Mesic hillside above Salem; August 1, 1977.

C-294 (3) *Adenanthera pavonina* L.; Jumbie Bead Tree; Fabaceae; Mesic forest above Salem; August 1, 1977.

C-295 (3) *Hymenaea courbaril* L.; Locust Tree; Fabaceae; Mesic forest above Salem; August 1, 1977.

C-296 (3) *Picrasma antillana* (Eggers) Urb.; Bitter

## Appendix

Ash; Simaroubaceae; Mesic forest above Salem; August 1, 1977.

C-297 (2) *Dieffenbachia seguine* (Jacq.) Schott; Dumb Cane; Araceae; Mesic hillside above Salem; August 1, 1977.

C-298 (5) *Pimenta racemosa* (Mill.) J. W. Moore; Bay Rum Tree; Myrtaceae; Mesic forest above Salem; August 1, 1977.

C-299 (2) *Gliricidia sepium* (Jacq.) Kunth ex Walp; Gorey Cedar; Fabaceae; Mesic hill above Salem; August 1, 1977.

C-300 (3) *Jasminum multiflorum* (Burm. f.) Andr.; Jasmine; Oleaceae; Mesic hillside, Olveston; August 1, 1977.

C-301 (5) *Hibiscus rosa-sinensis* L.; Hibiscus; Malvaceae; Woodlands, R. Marland's yard; August 1, 1977.

C-302 (4) *Terminalia catappa* L.; Beach Almond; Combretaceae; Lime Kiln Beach; August 1, 1977.

C-303 (3) *Momordica charantia* L.; Maiden Apple; Cucurbitaceae; Plymouth, roadside fence; August 2, 1977.

C-304 (3) *Passiflora foetida* L.; Baby Honeysuckle; Passifloraceae; White River Valley, xeric slope; August 2, 1977.

C-305 (3) *Dioscorea bulbifera* L.; Yam Pule; Dioscoreaceae; White River Valley, old field; August 2, 1977.

C-306 (3) *Exostema elliptica* Griseb.; Quina, Cinchona; Rubiaceae; Great Alp Falls area, mesic forest; August 2, 1977.

C-307 (5) *Nepeta cataria* L.; Catnip, Catmint; Lamiaceae; Dooryard garden, St. George's Hill; August 2, 1977.

C-308 (2) *Eupatorium villosum* Sw.; Sheep Mutton; Asteraceae; St. George's Hill, roadside; August 2, 1977.

C-309 (3) *Anethum graveolens* L.; Dillweed; Apiaceae; Dooryard garden, St. George's Hill; August 2, 1977.

C-310 (2) *Parthenium hysterophorus* L.; Wormwood, White Broomweed; Asteraceae; Dooryard garden, St. George's Hill; August 2, 1977.

C-311 (1) *Morinda citrifolia* L.; Chiddle Grape; Rubiaceae; Mesic slope, St. George's Hill; August 2, 1977.

C-312 (1) *Cynodon dactylon* (L.) Pers.; St. Augustine's Grass; Poaceae; St. George's Hill, roadside; August 2, 1977.

C-313 (1) *Helianthus hirsutus* Raf.; Sage; Asteraceae; St. George's Hill, roadside; August 2, 1977.

C-314 (1) *Chenopodium ambrosioides* L.; Worm-

C-315 (1) *Allium fistulosum* L.; Onion; Liliaceae; St. George's Hill, roadside; August 2, 1977. [preceded by: wood; Chenopodiaceae; St. George's Hill, roadside; August 2, 1977.]

C-316 (1) *Nicotiana tabacum* L.; Creole Tobacco; Solanaceae; St. George's Hill, dooryard garden; August 2, 1977.

C-317 (1) *Citharexylum spinosum* L.; Fiddlewood; Verbenaceae; St. George's Hill, dooryard garden; August 2, 1977.

C-318 (1) *Ambrosia artemisiifolia* L.; Tansy; Asteraceae; St. George's Hill, dooryard garden; August 2, 1977.

C-319 (1) *Emilia coccinea* (Sims) G. Don; Cabbage Bush; Asteraceae; St. George's Hill, dooryard garden; August 2, 1977.

C-320 (1) *Tropaeolum tuberosum* Ruiz & Pav.; Biscuit Plant; Tropaeolaceae; St. George's Hill, roadside; August 2, 1977.

C-321 (1) *Ruellia tuberosa* L.; Double Bit; Acanthaceae; St. George's Hill, mesic slope; August 2, 1977.

C-322 (1) *Catharanthus roseus* (L.) G. Don; Madagascar Periwinkle; Apocynaceae; Plymouth, dooryard; August 2, 1977.

C-323 (1) *Euterpe globosa* Gaertn.; Mountain Palm, Mountain Cabbage; Arecaceae; Chance Peak, rainforest jungle; August 2, 1977.

C-324 (1) *Castilla elastica* Sessé; Wild Rubber Tree; Moraceae; Chance Peak, mesic forested slope; August 2, 1977.

C-325 (1) *Crotalaria incana* L.; Rattle Bush, Shake Shake; Fabaceae; Plymouth, seaside; August 2, 1977.

C-326 (3) *Cucumis anguria* L.; West Indian Gherkin, Wild Cucumber; Cucurbitaceae; Vue Pointe Hotel, near old field; August 2, 1977.

C-327 (8) *Caesalpinia bonduc* (L.) Roxb.; Grey Nicker, Horse Nicker; Fabaceae; Trants Estate, xeric thicket; August 4, 1977.

C-328 (3) *Swietenia mahagoni* (L.) Jacq.; Mahogany; Meliaceae; Bugby Hole, edge of thicket; August 4, 1977.

C-329 (3) *Morisonia americana* L.; Rat Apple; Capparaceae; Trants Estate, xeric hillside; August 4, 1977.

C-330 (2) *Ficus stahlii* Warb.; Moraceae; Paradise, near old estate house site; August 4, 1977.

C-331 (2) *Piscidia piscipula* (L.) Sarg.; Dogwood; Fabaceae; Harris Village, mesic roadside thicket; August 4, 1977.

C-332 (2) *Piscidia carthagenensis* Jacq.; Fish Poi-

## Appendix

son Tree; Fabaceae; Harris Village, mesic roadside thicket; August 4, 1977.

C-333 (1) *Emilia sonchifolia* (L.) DC.; Rabbit Food; Asteraceae; Cudjoehead Village, mesic roadside; August 4, 1977.

C-334 (1) *Abrus precatorius* L.; Jumbie Beads, Crab's Eyes; Fabaceae; Cudjoehead Village, roadside thicket; August 4, 1977.

C-335 (2) *Centrosema virginianum* (L.) Benth.; Wild Pea; Fabaceae; Cudjoehead Village, mesic slope; August 4, 1977.

C-336 (2) *Ruellia tuberosa* L.; Double Bit; Acanthaceae; Cudjoehead Village, mesic roadside ditch; August 4, 1977.

C-337 (1) *Lantana camara* L.; Red Sage; Verbenaceae; Cudjoehead Village, mesic roadside ditch; August 4, 1977.

C-338 (1) *Galactia filiformis* Benth.; Strongback; Fabaceae; Cudjoehead Village, mesic ghaut; August 4, 1977.

C-339 (1) *Pseudelephantopus spicatus* (B. Juss. ex Aubl.) C. F. Baker; Bull Tongue, Cattle Tongue Bush; Asteraceae; Cudjoehead Village, roadside; August 4, 1977.

C-340 (3) *Wedelia trilobata* (L.) Hitchc.; Carpet Daisy; Asteraceae; Cudjoehead Village, mesic roadside; August 4, 1977.

C-341 (1) *Tamarindus indica* L.; Tamarind Bush; Fabaceae; Cudjoehead Village, mesic slope; August 4, 1977.

C-342 (3) *Commelina elegans* Kunth.; White Weed; Commelinaceae; Cudjoehead Village, mesic roadside; August 4, 1977.

C-343 (1) *Macfadyena unguis-cati* (L.) A. Gentry; Cat's Claw; Bignoniaceae; Cudjoehead Village, mesic roadside; August 4, 1977.

C-344 (1) *Momordica charantia* L.; Pom Cooly; Cucurbitaceae; Cudjoehead Village, mesic roadside; August 4, 1977.

C-345 (3) *Erythroxylum havanense* Jacq.; Lionwood, Vinette, Barberry; Erythroxylaceae; Xeric hilltop above Rendezvous Bay; August 8, 1977.

C-346 (2) *Crotalaria incana* L.; Rattle Bush, Shake Shake; Fabaceae; Plymouth, seaside at edge of road; August 8, 1977.

C-347 (1) *Justicia pectoralis* Jacq.; Garden Balsam; Acanthaceae; Public market, Plymouth; August 8, 1977.

C-348 (1) *Croton flocculosus* Geisl.; Bitter Balsam; Euphorbiaceae; Rocky ghaut near Olveston; August 8, 1977.

C-349 (3) *Annona squamosa* L.; Sugar Apple; Annonaceae; Olveston, mesic hillside; August 8, 1977.

C-350 (1) *Pithecellobium saman* (Jacq.) Benth.; Rainfall Tree; Fabaceae; Olveston, mesic hillside; August 8, 1977.

C-351 (1) *Pimenta racemosa* (Mill.) J. W. Moore; Bay Rum Tree, Celemon Bush; Myrtaceae; Olveston, mesic hillside; August 8, 1977.

C-352 (1) *Bursera simaruba* (L.) Sarg.; Gum Bush, Gum Tree; Burseraceae; Olveston, mesic hillside; August 8, 1977.

C-353 (1) *Chiococca alba* (L.) Hitchc.; Davis Root, David's Root; Rubiaceae; Olveston, mesic hillside forest; August 8, 1977.

C-354 (3) *Hibiscus sabdariffa* L.; Sorrel, Roselle; Malvaceae; Salem, dooryard garden of Alfred Payne; December 19, 1978.

C-355 (1) *Cajanus cajan* (L.) Huth; Pigeon Pea; Fabaceae; Salem, dooryard garden of Alfred Payne; December 19, 1978.

C-356 (3) *Psychotria domingensis* Jacq.; Bois Cabrit; Rubiaceae; Olveston, mesic hillside; December 21, 1978.

C-357 (3) *Hura crepitans* L.; Sandbox Tree; Euphorbiaceae; Olveston, mesic hillside; December 24, 1978.

C-358 (3) *Eupatorium odoratum* L.; Christmas Bush; Asteraceae; Centre Hills, bamboo forest, near footpath; December 26, 1978.

C-359 (3) *Palicourea crocea* (Sw.) Roem. and Schult.; Red Palicourea; Rubiaceae; Centre Hills, bamboo forest, mesic slope; December 26, 1978.

C-360 (3) *Elephantopus mollis* H. B. K.; Zou Mouton; Asteraceae; Centre Hills, bamboo forest, near footpath; December 26, 1978.

C-361 (7) *Sauvagesia erecta* L.; Bush Tea, Mountain Tea; Ochnaceae; Centre Hills, bamboo forest, near footpath; December 26, 1978.

C-362 (2) *Thunbergia alata* Boj. ex Sims; Golden Bells; Acanthaceae; Centre Hills, bamboo forest, near footpath; December 26, 1978.

C-363 (3) *Gleichenia dichotoma* (Thunb.) Hook.; Polypodiaceae; Centre Hills, bamboo forest, mesic slope; December 26, 1978.

C-364 (1) *Charianthus purpureus* D. Don; Wassard; Melastomaceae; Centre Hills, bamboo forest, mesic slope; December 26, 1978.

C-365 (2) *Miconia prasina* (Sw.) DC.; Camasey; Melastomaceae; Near bottom of Chance Peak, mesic slope, December 28, 1978.

C-366 (3) *Flemingia strobilifera* (L.) Aiton f.; Wild Hops; Fabaceae; Near bottom of Chance Peak, mesic slope; December 28, 1978.

## Appendix

C-367 (2) *Cephaelis mucosa* Sw.; Ipéca Bâtard, Bois Marguerite; Rubiaceae; Chance Peak, rainforest, 61 m; December 28, 1978.

C-368 (2) *Cephaelis swartzii* DC.; Ipéca Bâtard, Bois Marguerite; Rubiaceae; Chance Peak, rainforest, 610 m; December 28, 1978.

C-369 (5) *Phyllanthus mimosoides* Sw.; Iron Bark; Euphorbiaceae; Chance Peak, rainforest, 762 m; December 28, 1978.

C-370 (4) *Cranichus mucosa* Sw.; Orchidaceae; Chance Peak, rainforest, 823 m; December 28, 1978.

C-371 (2) *Habenaria alata* Hook.; Orchidaceae; Chance Peak, rainforest, 884 m; December 28, 1978.

C-372 (2) *Antigonon leptopus* Hook. & Arn.; Coralita; Polygonaceae; The Groves, roadside; December 28, 1978.

C-373 (4) *Talinum triangulare* (Jacq.) Willd.; Pourpier Grand Bois; Portulacaceae; The Groves, roadside; December 28, 1978.

C-374 (2) *Cissus sicyoides* L.; Skip Rope, Scratch Wythe, Poison Wythe; Vitaceae; Cork Hill, mesic slope; December 28, 1978.

C-375 (1) *Cassia occidentalis* L.; Wild Coffee; Fabaceae; Olveston, edge of mesic thicket; January 2, 1979.

C-376 (2) *Dicliptera martinicensis* (Jacq.) Juss.; Acanthaceae; Olveston, thorn forest; January 2, 1979.

C-377 (3) *Tecoma stans* (L.) Juss. ex Kunth.; Elderbush; Bignoniaceae; Mesic hillside above Salem; January 2, 1979.

C-378 (3) *Vernonia cinerea* (L.) Less.; Measle Bush; Asteraceae; Mesic hillside on Cavalla Hill; January 2, 1979.

# References Cited

Bayley, I. 1949. The Bush-teas of Barbados. *The Journal of the Barbados Museum and Historical Society* 16(3): 103–12.

Beard, J. S. 1949. *The natural vegetation of the Windward and Leeward Islands.* Oxford: The Clarendon Press.

Britton, N. L., and C. F. Millspaugh. 1962. *The Bahama flora.* New York: Published for the New York Botanical Garden by Hafner Publishing Co.

Brussell, D. E. 1977. That great Christmas mistletoe. *Illinois Magazine* 16(10): 57–58.

Coats, A. M. 1969. *The plant hunters.* New York: McGraw-Hill Book Co.

Correll, D. S., and H. B. Correll. 1982. *Flora of the Bahama Archipelago (including the Turks and Caicos Islands).* Vaduz: J. Cramer.

Davis, W. 1985. *The serpent and the rainbow.* New York: Simon and Schuster.

Davis, W. 1988. *Passage of darkness: The ethnobiology of the Haitian Zombie.* Chapel Hill: The University of North Carolina Press.

Dobbin, Jay D. 1986. *The jombee dance of Montserrat.* Columbus: Ohio State University Press.

Duberry, A. 1973. Folk medicines of Montserrat and their uses. Plymouth, Montserrat: Files of the author.

Fawcett, W. and A. B. Rendle. 1910. *Flora of Jamaica.* London: Her Majesty's Stationery Office.

Fergus, H. A. 1975. *History of Alliouagana. A short history of Montserrat.* Plymouth, Montserrat: Montserrat Printery.

Fritzmaurice, L. W. 1953. *The vomiting sickness of Jamaica.* The West Indian Medical Journal. 2(2): 93–124.

Gooding, E. G. B., A. R. Loveless, and G. R. Proctor. 1965. *Flora of Barbados.* London: Her Majesty's Stationery Office.

Grisebach, A. H. R. 1864. *Flora of the British West Indian Islands.* London: Lovell Reeve & Company.

Harlow, V. T. 1924. *Colonizing expeditions to the West Indies and Guiana 1623–1667.* 2d ser., vol. 56. London: Hakluyt Society Publications.

## References Cited

Harris, D. R. 1963. Plants, animals, and man in the Outer Leeward Islands, West Indies: An ecological study of Antigua, Barbuda, and Anguilla. Ph.D. diss. University of California, Berkeley.

Herman, E. 1973. *Montserrat cookbook.* Woodstock, Ontario: Woodstock Print & Litho Ltd.

Hodge, W. H., and D. Taylor. 1957. The ethnobotany of the Island Caribs of Dominica. *Webbia* 12(2): 513–644.

Howard, R. A. 1961. *Why Montserrat? Leewards.* Ed. John Brown. Barbados: The Advocate Company Ltd.

Howard, R. A. 1974–1989. *Flora of the Lesser Antilles (Leeward and Windward Islands).* 6 vols. Jamaica Plain, Massachusetts: Arnold Arboretum, Harvard University.

Irish, J. A. G. 1973. *Alliouagana in focus.* Plymouth, Montserrat: Montserrat Printery.

Jesse, Rev. C. 1966. An houre glasse of Indian newes: A record of the settlement in St. Lucia in 1605. *Caribbean Quarterly* 12(1): 46–47.

Lee, C. 1980. Schistosomiasis in Montserrat, a prevalence survey in the Leeward Islands. Master's thesis. University of London, London.

Lee-Huang, S., Philip L. Huang, Peter L. Nara, Hao-Chia Chen, Hsiang-Fu Kung, Peter Huang, Henry I. Huang, and Paul L. Huang 1990. *MAP 30: a new inhibitor of HIV-1 infection and replication.* Federation of European Biochemical Societies. 272(1): 12–18.

Little, E. L., and F. H. Wadsworth. 1964. *Common trees of Puerto Rico and the Virgin Islands.* U. S. Department of Agriculture. Agriculture Handbook No. 29. Washington: Government Printing Office.

Lopinot, N.H., and D.E. Brussell. 1982. Assessing uncarbonized seeds from open-air sites in mesic environments: An example from southern Illinois. *Journal of Archaeological Science* 9(1): 95–108.

Merrill, E. D. 1954. *The botany of Cook's voyages.* Waltham, Massachusetts: Chronica Botanica Company.

Montserrat National Trust. 1976. *History of Montserrat.* Plymouth, Montserrat: The Montserrat National Trust Museum.

Morton, J. F. 1981. *Atlas of Medicinal Plants of Middle America: Bahamas to Yucatan.* Springfield, Illinois: Charles C. Thomas.

Morton, J. F. 1982. *Plants poisonous to people in Florida and other warm areas.* 2d. ed. Coral Gables, Florida: J. F. Morton.

*New Rand McNally college world atlas, The.* 1983. Chicago: Rand McNally & Company.

Oliver, V. L. 1910–1919. *Caribbeana: Being miscellaneous papers relating to the history, genealogy, topography, and antiquities of the British West Indies.* 6 vols. London: Mitchell, Hughes & Clark.

Pulsipher, L. M. 1977. The cultural landscape of Montserrat, West Indies, in the seventeenth century: Early environmental consequences of British colonial development. Ph.D. diss. Southern Illinois University, Carbondale.

Rouse, I. 1946. *The Carib Indian. Handbook of South American Indians.* Washington: Bureau of American Ethnology, Smithsonian Institution.

Schultes, R. E., and A. Hofmann. 1980. *The botany and chemistry of hallucinogens.* Springfield, Illinois: Charles C. Thomas.

Shafer, J. A. 1907. Report on a visit to the island of Montserrat. *Journal of the New York Botanical Garden* 8(88): 81–86.

Simpson, B. B., and M. Conner-Ogorzaly. 1986. *Economic botany: Plants in our world.* New York: McGraw-Hill Book Company.

Smith, J. 1630. *The true travels, adventures, and observations of Captain John Smith from 1593–1629.* London: Printed by Thomas Slater.

Taylor, D. 1938. *The Caribs of Dominica.* Washington: Bureau of American Ethnology, Smithsonian Institution.

Ugent, D., S. Pozorski, and T. Pozorski. 1984. New evidence for ancient cultivation of *Canna edulis* in Peru. *Economic Botany* 38(4): 417–32.

Watters, D. R. 1980. Transect surveying and prehistoric site locations on Barbuda and Montserrat, Leeward Islands, West Indies. Ph.D. diss. The University of Pittsburgh.

Williams, R. O., and R. O. Williams, Jr. 1951. *The useful and ornamental plants in Trinidad and Tobago.* Port of Spain: Guardian Commercial Printery.

# Index of Common Names

Acerola Cherry, 96, 149
Agouman, 143
Akee, 120, 121, 146
Alliouagana, 12
Aloe, 93, 151
Angel's Trumpet, 84, 125
Anise, 24
Annatto, 41, 73
Aphrodite's Apple, 84, 122, 150
Applebush, 23
Areca Nut, 32
Arrowroot, 98
Avocado, 93

Baby Honeysuckle, 108, 138, 141, 152
Balisier, 81, 104, 147
Balsam, 40, 56, 91, 130, 139
Bamboo, 110, 148
Banana, 91, 101, 104, 105
Banyan, 150
Barbados Cherry, 96, 139, 149
Barberry, 154

Batroot, 55, 109, 138, 141
Bay Rum Tree, 105, 152, 155
Beach Almond, 52, 152
Beach Morning Glory, 53, 75, 135
Beggartick, 37
Bermuda Grass, 110
Betel Palm, 32
Big Chaney Bush, 29, 70
Bird Lime, 63
Bird of Paradise, 148
Biscuit Plant, 129, 153
Bitter Aloe, 93, 151
Bitter Ash, 26, 123, 151
Bitter Balsam, 15, 56, 154
Bitter Bark, 123
Bitter Bush, 26
Black Apple, 55, 76, 142, 148
Black Birch, 105
Black Candlewood, 118
Black Fig, 150
Black Torch, 92
Bluebell, 68

Blue Flower, 130
Bobena, 40
Body Cutter, 58, 138
Bois Bande, 128
Bois Cabrit, 155
Bois Gleau, 151
Bois Marguerite, 147, 151, 156
Bois Puce, 136
Bois Violin, 151
Boulangie, 126
Bowstring Hemp, 95
Bread and Cheese, 88, 149
Breadfruit, 80, 100, 142
Briny Roots, 117
Broad Leaf Nettle, 129
Bull Tongue, 39, 154
Burma Rosewood, 89, 149
Burry Bark, 96, 146
Bush Tarragon, 136
Bush Tea, 155
Buttercup, 41
Butterfly Flower, 141

## Index of Common Names

Cabbage Bush, 38, 153
Cabbage Palm, 36, 143
Cacao, 126, 137
Calabash, 40, 72, 138
Calalu, 30
Camasey, 98, 155
Caper Tree, 149
Carpet Daisy, 40, 154
Casha, 64, 135
Cashaw, 65
Cashee, 88
Cashew, 17
Cashew Tree, 138
Cassava, 11, 58, 59, 60, 61, 139, 140
Castor Nut, 62
Castor Oil Plant, 62
Catmint, 90, 137, 139, 152
Catnip, 90, 137, 139, 152
Cat's Claw, 41, 138, 154
Cattle Tongue Bush, 39, 154
Celemon Bush, 105, 155
Chaney Bush, 29
Chayote, 55
Cherimoya, 20
Chiddle Grape, 114, 118, 152
Chigger Nut, 44
Chivel, 93
Chorita, 131
Christmas Bush, 18, 38, 53, 118, 146, 155

Christmas Flower, 57
Christmas Wreath, 53
Christophine, 55, 76, 143
Ciguatera, 22, 41, 64, 65
Cinchona, 152
Cinnamon, 92, 149
Cinnamon Tree, 105
Clammy Cherry, 43, 44, 135
Climbing Tongue Fern, 114, 140
Cochineal Cactus, 47, 48
Cocoa, 126, 127, 128
Cocoa Tree, 126, 151
Coconut, 32, 33, 34, 35, 71
Coffee Tree, 117
Coot Weed, 151
Coralita, 114, 135, 156
Coratoe, 15, 16
Coretor, 51, 95, 144
Corn, 113
Corn Grass, 113, 142, 147
Corossol Marron, 151
Cow Itch, 63, 141
Cow Itch Vine, 86
Cow's Tongue, 43, 140
Crab's Eyes, 63, 154
Crackers, 26, 27, 29, 148
Creole Tobacco, 126, 153
Cum Bush, 45
Cupid's Paint Brush, 38
Cure for All, 39

Currant Tree, 128
Cusha, 64
Custard Apple, 22
Cypre, 42

Dasheen, 28, 142
Date Palm, 36, 37
David's Root, 117, 137, 139, 150, 151, 155
Davis Root, 117, 155
Devil's Grass, 110
Devil's Horse Whip, 16, 143
Devil's Tail, 130, 143
Dildo, 46
Dill, 24
Dillweed, 24, 152
Dr. Dyette, 25
Dog Apple, 20
Dogwood, 78, 87, 138, 146, 153
Dotma, 29
Double Bit, 15, 143, 153, 154
Dumb Cane, 28, 152
Duppy Basil, 90
Duppy Needles, 37, 145
Dutchgrass, 110, 145
Dyewood, 85

Eddee, 28, 142
Eddo, 28
Eggplant, 126

## Index of Common Names

Elder, 43, 50, 143, 145
Elder Bush, 41, 109, 140, 156
English Plantain, 110
Eucalyptus, 105, 141
Everyday Flower, 25, 137
Evil Tree, 103, 148
Eyebright, 56, 130, 146, 151

Fever Grass, 110, 139
Fiddlewood, 129, 153
Fish Poison Tree, 87, 153
Fitweed, 24
Flamboyant, 150
Fraise, 116
French Body Cutter, 58, 138, 149
Friendego, 64, 65, 144

Garden Balsam, 154
Garlic, 93
Genip, 117, 122, 141, 143
Giant Tree, 109
Ginger, 131, 148
Ginger Thomas, 41
Goat Grass, 136
Goat Hahn, 118, 145
Golden Apple, 19, 149
Golden Bells, 135, 155
Golden Shower, 67
Gooseberry Tree, 62, 149
Gorey Cedar, 85, 145, 152

Governor's Plum, 19, 89, 146, 149
Granada, 116
Graveyard Grass, 40
Grey Nicker, 66, 77, 153
Groundnut, 66, 137, 139
Gru Gru Palm, 30, 31
Guava, 106, 139, 150, 151
Gully Plum, 20
Gum Bark, 45, 137
Gum Bush, 45, 155
Gummy Lingo, 45
Gum Tree, 45, 137, 140, 155

Hardgrass, 110
Hard Leaf, 26
Heart Bush, 122, 142
Hibiscus, 79, 96, 152
Hog Apple, 118
Hog Plum, 20, 149
Hog Vine, 53, 139
Hogwood, 98, 150
Holly, 18
Honduras Mahogany, 100, 146
Honey Bush, 151
Horse Nicker, 66, 153
Horseradish Tree, 103, 148
Hotpepper Bush, 124, 141
Hug Me Close, 16

Indian Almond, 52

Indian Root, 36
Indigo, 137
Indigo Weed, 85, 137
Inflammation Bush, 109, 136, 140
Ink Balls, 143
Ipéca Bâtard, 147, 151, 156
Iron Bark, 156

Jackfruit, 101, 149
Jacob's Ladder, 56, 142
Jamaica Plum, 20
Jambolan, 107
Jasmine, 106, 152
Java Plum, 89, 106, 146
John Crow Beads, 63
Jumbie Apple, 20
Jumbie Bead, 63, 137, 148, 154
Jumbie Bead Tree, 65, 68, 151
Jumbie Berry, 150
Jumbie Tomato, 126

Khus Khus, 113
Kidney Bean, 86
Kumquat, 119

Lancewood, 92
Langford Cabbage, 43
Lantana, 129
Leaf of Life, 53
Leley, 43, 141

## Index of Common Names

Lemon Grass, 110, 139
Liane Percée, 147
Lignum Vitae, 131, 149
Lilac, 100, 138, 141
Lima Bean, 86
Lime, 13, 119, 148
Lionwood, 154
Lipstick Plant, 41, 149
Lizard's Food, 54
Locust Tree, 85, 141, 151
Logwood, 85, 149
Long Grass, 113, 136, 138, 144
Loofah, 54
Lord Lovington, 89, 143
Love, 92
Love Bush, 53, 143
Lovegrass, 110
Love Leaf, 53
Love Vine, 38, 92, 149
Luckynut, 26, 144
Luckyseed, 26, 144

Macadamia, 115, 149
Madagascar Periwinkle, 25, 153
Mahogany, 100, 101, 153
Maiden Apple, 54, 75, 152
Malay Apple, 107
Malimbé, 136
Mammy, 51, 151
Mammy Apple, 51, 74, 136, 140

Man Better Man, 16, 130, 151
Manchineel, 57, 143
Mandarin, 119
Mango, 18, 19, 142, 151
Mangrove, 9
Mansipote, 51
Man Tree, 37, 137
Many Roots, 15
Marjoram, 90
Masquerade Whip, 131
Mauve Savane, 150
Mawby, 116
Maypole, 15
Measle Bush, 40, 129, 144, 156
Milk Bush, 26
Milktree, 25
Milky, 25, 144
Mistletoe, 95
Money Bush, 67, 138
Money Money Bush, 67, 151
Money Money Tree, 67, 141
Money Plant, 67, 145
Monkey Apple, 20
Monkey Gun, 15
Monkey Nut, 66, 139
Monster, 29
Mosquito Bush, 90
Mountain Cabbage, 36, 153
Mountain Palm, 36, 153
Mountain Porridge, 49, 146

Mountain Tea, 155
Muguet Bleu, 147

Natal Plum, 24, 148
Neem, 99
Nettle, 145
Nightblooming Cereus, 46
Nora Bush, 137
No Mammy, 45, 95, 140

Oil Nut, 62
Oleander, 25, 148
Onion, 153
Oregano, 91, 130, 139, 145
Overlooker, 66

Painkiller, 118, 124, 138, 150
Papaya, 50
Passion Apple, 122, 150
Passionfruit, 108, 143
Pasture Sage, 40, 145
Pawpaw, 50
Peanut, 66, 117, 137, 139
Peas, 66
Pencil Plant, 57, 144
Peppermint, 89, 139
Pigeonberry, 42, 126, 136
Pigeon Pea, 66, 136, 155
Pine, 44
Pineapple, 44

## Index of Common Names

Pipe Organ Cactus, 46
Plantain, 104
Poinsettia, 57
Poison Ivy, 18, 19
Poison Tree, 63
Poison Wythe, 131, 137, 156
Pom Cooly, 54, 135, 154
Pomegranate, 116, 148
Pond Apple, 20
Pop Bush, 143
Pourpier Grand Bois, 155
Prickle Pear, 48
Prickly Palm, 32
Pung Cooly, 54
Purslane, 115, 139
Pursly, 115, 139
Pusly, 115, 141

Quina, 152

Rabbit Food, 38, 142, 154
Rainfall Tree, 85, 88, 138, 155
Raintree, 88, 141
Ramgoat Bush, 24, 117, 136
Raspberry, 116, 145
Rat Apple, 149, 153
Rat Bean, 149
Rattle Bush, 143, 153, 154
Red Cedar, 98, 99, 138, 140
Red Frangipani, 26, 69, 144

Red Grape, 114
Redhead, 36
Red Iron Bark, 62, 145
Red Manjack, 43
Red Milkweed, 56
Red Palicourea, 155
Red Rodwood, 105, 142
Red Sage, 129, 140, 154
Red Trouble, 126
Red Trubba, 126
Redwood, 100, 142
Right Wythe, 41
Rock Balsam, 56, 137, 146, 150
Rock Bush, 109, 140
Rope Bush, 96, 136
Rope Plant, 97, 150
Rose Apple, 106, 107, 148
Roselle, 97, 148, 155
Rosemary, 91, 139
Rose Plum, 106, 139
Rubber Hedge Plant, 57

Sage, 38, 39, 140, 146, 152
Sage Bush, 43
St. Augustine's Grass, 110, 152
Saize Bush, 43, 138
Sandbox Tree, 57, 77, 155
Sapodilla, 122
Sarsaparilla, 29, 145, 146
Satinwood, 120

Scotch Attorney, 51
Scratch Wythe, 131, 156
Seagrape, 83, 114, 151
Sea Island Cotton, 96, 97, 150
Sea Lavender, 44
Seaside Mahoe, 98
Seaside Sage, 56
Sedge, 137
Seepwood, 42, 144
Seguine Couleuvre, 147
Shake Shake, 143, 153, 154
Shame Bush, 86, 138
Sheep Grass, 137
Sheep Mutton, 38, 152
Shell Ginger, 131
Shell Plant, 131, 144
Shoemaker Bark, 95, 144
Shrove Tuesday, 68
Sierra Palm, 36
Silkcotton, 42, 74, 145
Sintibibi, 93, 94, 151
Sisal, 16
Skip Rope, 95, 131, 146, 156
Skip Vine, 95
Sleepy, 86
Snake Bush, 114
Snakeweed, 145
Snakewood, 102, 123, 145
Soldier Bush, 44
Soldier Rod, 16

## Index of Common Names

Soldierwood, 116
Sorrel, 80, 97, 142, 148, 155
Sour Orange, 119
Soursop, 21, 22, 141
Southernwood, 93, 139
Spanish Lime, 122
Spanish Needles, 37, 145
Spanish Oak, 86, 142, 151
Spearmint, 90
Spinach, 17, 136, 148
Star of Bethlehem, 49, 137
Stinking Bush, 67, 143, 151
Strongback, 42, 68, 86, 137, 141, 145, 154
Strongbark, 42
Strong Man's Weed, 109
Strongwood, 131
Sugar Apple, 23, 138, 154
Sugarcane, 13, 82, 111, 112
Sweet Acacia, 64
Sweet Balsam, 56, 141
Sweet Grass, 110, 113, 141
Sweet Lime, 120, 141
Sweet Marjoram, 90, 139
Sweet Pod, 85
Sweet Potato, 29, 52
Sweetsop, 23, 146
Sweet Tamarind, 88
Sweetwood, 92, 139, 142, 145
Swivel Sweet, 119, 140

Taman, 89
Tamarind, 89, 138
Tamarind Bush, 89, 154
Tangerine, 119, 140
Tansy, 39, 139, 153
Taro, 28
Tarragon, 37
Teak, 130
Thorn Palm, 32
Thyme, 91, 150
Tickleweed, 115, 147
Ticky Thyme, 91
Ti Codinde, 148
Tobacco, 13
Toiseloisemoise, 49, 146
Toloma, 49, 150
Toloma Food, 49, 146
Tomato, 125
Tom Tah, 123, 144
Torchwood, 118
Trumpeter, 148
Trumpeter Tree, 102
Trumpet Tree, 148
Turk's Cap, 47
Turk's Head, 47
Twelve O'clock, 25, 98, 137, 138, 142, 144

Vervain, 130
Vete Vere Grass, 113

Vine Nettle, 63, 141
Vinette, 154

Wanger Bush, 125
Wassard, 136, 155
Watchman, 66
Water Grass, 52
Welsh Onion, 93, 146
West Indian Gherkin, 54, 153
West Indian Plum, 20, 139
White Bean, 86
White Broomweed, 152
White Castor Nut, 62, 140
White Cedar, 41, 146
White Deal, 123
White Eggplant, 127
White Frangipani, 25, 144
White Frenchweed, 52, 141
White Manjack, 43
White Weed, 52, 154
Whitewood, 41
Whorewine, 130, 136, 145
Wild Blue Vine, 68
Wild Coffee, 67, 156
Wild Cucumber, 54, 153
Wild Frangipani, 25, 146
Wild Hops, 155
Wild Ipecacuanha, 36, 136
Wild Licorice, 63
Wild Mammy Apple, 51, 136, 144

## Index of Common Names

Wild Mansipote, 51
Wild Okra, 97
Wild Pea, 68, 142, 144, 154
Wild Rubber, 81
Wild Rubber Tree, 102, 140, 147, 153
Wild Sarsaparilla, 29, 123, 150
Wild Soursop, 21, 145
Wild Strawberry, 46, 146
Wild Tamarind, 67, 87, 150
Wild Tarragon, 38, 143

Wonga, 124, 146
Worm Grass, 95
Worm Oil Tree, 51, 141
Wormseed Weed, 51, 135
Wormwood, 39, 51, 135, 152
Worwine, 130

Yam, 55
Yam Pule, 152

Yellow Balsam, 56
Yellow Dodder, 92
Yellow Harklis, 120
Yellow Hercules, 120
Yellow Mombin, 20
Yellow Prickle, 120, 140
Yellow Sapote, 122, 150

Zou Mouton, 155

# Index of Scientific Names

*Abrus precatorius*, 63, 137, 148, 154
*Abutilon hirtum*, 96, 146
*Abutilon indicum*, 96, 136
*Acacia*, 9, 12
*Acacia farnesiana*, 64
*Acacia macracantha*, 64, 135
*Acacia tortuosa*, 65, 144
*Acalypha virgata*, 148
Acanthaceae, 15, 135, 143, 153, 154, 155, 156
*Achyranthes indica*, 16, 143
*Acrocomia aculeata*, 30, 31
*Acrocomia media*, 32
*Adelia ricinella*, 55, 138, 141
*Adenanthera pavonina*, 65, 151
Agavaceae, 15
*Agave beauleriana*, 15, 16
*Agave franzosini*, 15
*Agave sisalana*, 16
*Allium fistulosum*, 93, 146, 153
*Allium sativum*, 93
*Aloe vera*, 93, 94, 151

*Alpinia nutans*, 131
*Alpinia zerumbet*, 131, 144
Amaranthaceae, 16, 136, 143, 148
*Amaranthus dubius*, 17, 136, 148
*Ambrosia artemisiifolia*, 153
*Amyris elemifera*, 118
Anacardiaceae, 17, 138, 139, 142, 149, 151
*Anacardium occidentale*, 17, 138
*Ananas comosus*, 12, 44
*Andropogon bicornis*, 137
*Anethum graveolens*, 24, 152
Annonaceae, 20, 138, 141, 145, 146, 151, 154
*Annona cherimola*, 20
*Annona glabra*, 20
*Annona montana*, 21, 145
*Annona muricata*, 21, 22, 141
*Annona reticulata*, 22
*Annona squamosa*, 23, 138, 146, 154
*Anthurium cordatum*, 147
*Anthurium grandifolium*, 26, 27

*Anthurium lanceolatum*, 147
*Anthurium willdenowii*, 147
*Antigonon leptopus*, 114, 135, 156
Apiaceae, 24, 136, 152
Apocynaceae, 24, 137, 144, 146, 148, 153
Araceae, 26, 142, 146, 147, 148, 152
*Arachis hypogaea*, 12, 66, 117, 137, 139
*Areca catechu*, 32
Arecaceae, 30, 143, 153
*Artemisia dracunculoides*, 37, 136
*Artocarpus altilis*, 80, 100, 142
*Artocarpus communis*, 100
*Artocarpus heterophyllus*, 101, 149
Asclepiadaceae, 36, 136, 137
*Asclepias curassavica*, 36, 136
Asteraceae, 37, 136, 140, 142, 143, 145, 146, 152, 153, 154, 155, 156

Balsaminaceae, 40
*Bambusa vulgaris*, 110, 147

## Index of Scientific Names

*Beilschmiedia pendula*, 9
*Bernardia dichotoma*, 137
*Beslaria lutea*, 151
*Bidens pilosa*, 37, 145
Bignoniaceae, 40, 138, 144, 146, 154, 156
Bixaceae, 41, 149
*Bixa orellana*, 11, 41, 73, 149
*Blighia sapida*, 120, 121, 146
Bombacaceae, 42, 145
Boraginaceae, 42, 135, 137, 140, 141, 144
*Bourreria succulenta*, 42
Bromeliaceae, 44, 150
*Bryophyllum pinnatum*, 53, 143
Burseraceae, 45, 137, 140, 155
*Bursera simaruba*, 9, 45, 137, 140, 155
*Byrsonima crassifolia*, 9
*Byrsonima spicata*, 95, 144

Cactaceae, 46, 146
*Cactus intortus*, 47
*Caesalpinia bonduc*, 66, 77, 153
*Cajanus cajan*, 66, 136, 155
*Calophyllum antillanum*, 9
Campanulaceae, 49, 137
*Campyloneuron phyllitidis*, 147
*Canavalia ensiformis*, 66
Cannaceae, 49, 146, 150
*Canna edulis*, 49, 146

*Canna generalis*, 150
Capparaceae, 149, 153
*Capparis flexuosa*, 149
Caprifoliaceae, 50, 143
*Capsicum annuum*, 12, 124, 141
Caricaceae, 50
*Carica papaya*, 12, 50
*Carissa macrocarpa*, 24, 148
*Cassia bicapsularis*, 67, 141, 145, 151
*Cassia fistula*, 67, 144
*Cassia glandulosa var. swartzii*, 67, 150
*Cassia obtusifolia*, 67, 138
*Cassia occidentalis*, 67, 143, 151, 156
*Cassia planisiliqua*, 68, 144
*Cassytha filiformis*, 38, 92, 149
*Castilla elastica*, 81, 102, 140, 147, 153
*Catharanthus roseus*, 25, 137, 153
*Cecropia peltata*, 9, 102, 148
*Cedrela mexicana*, 9, 98, 99, 138, 140
*Ceiba pentandra*, 9, 12, 42, 74, 145
*Centrosema virginianum*, 68, 142, 154
*Cephaelis axillaris*, 151
*Cephaelis mucosa*, 156
*Cephaelis swartzii*, 147, 156
*Cephalocereus royenii*, 9, 46
*Charianthus alpinus*, 9
*Charianthus purpureus*, 136, 155
Chenopodiaceae, 51, 135, 141, 153
*Chenopodium ambrosioides*, 51, 135, 141, 152

*Chiococca alba*, 117, 123, 137, 139, 150, 151, 155
*Cinnamomum zeylanicum*, 92, 149
*Cissus sicyoides*, 131, 137, 146, 156
*Citharexylum spinosum*, 129, 153
*Citrus aurantifolia*, 119, 148
*Citrus aurantium*, 119, 140
*Citrus reticulata*, 119, 140
*Clusia alba*, 10
Clusiaceae, 51, 136, 140, 144
*Clusia rosea*, 9, 51, 136, 144
*Coccoloba swartzii*, 114
*Coccoloba uvifera*, 83, 114, 151
*Coccoloba venosa*, 114
*Cocos nucifera*, 32, 33, 71
*Codiaeum variegatum*, 147
*Coffea arabica*, 117
*Colocasia esculenta*, 28, 142
*Colubrina elliptica*, 116
Combretaceae, 52, 152
Commelinaceae, 52, 141, 154
*Commelina diffusa*, 52, 141
*Commelina elegans*, 52, 154
*Comocladia dodonaea*, 18
Convolvulaceae, 52, 135, 137, 139, 146
*Cordia alliodora*, 42, 144
*Cordia collococca*, 43
*Cordia curassavica*, 43, 138, 140
*Cordia nitida*, 43, 141

# Index of Scientific Names

*Cordia obliqua*, 43, 44, 135
*Cranichus mucosa*, 156
Crassulaceae, 53, 143
*Crescentia cujete*, 12, 40, 72, 138
*Crotalaria incana*, 143, 153, 154
*Crotalaria lotifolia*, 151
*Crotalaria quinquefolia*, 143
*Croton*, 9
*Croton balsamiferum*, 56
*Croton flavens*, 56, 137, 141, 146, 150
*Croton flocculosus*, 56, 154
*Cucumis anguria*, 54, 153
Cucurbitaceae, 54, 135, 143, 152, 153, 154
*Cyathea arborea*, 9
*Cymbopogon citratus*, 110, 139
*Cynodon dactylon*, 110, 145, 152
Cyperaceae, 137, 147
*Cyperus ligularis*, 10

*Dacryodes*, 9
*Dacryodes excelsa*, 9
*Datura innoxia*, 124, 138, 146
*Datura metel*, 124
*Datura stramonium*, 125
*Datura suaveolens*, 84, 125
*Delonix regia*, 150
*Dicliptera martinicensis*, 156
*Didymopanax attenuatum*, 9
*Dieffenbachia seguine*, 28, 152

*Digitalis purpurea*, 122, 142
*Digitaria insularis*, 113
*Dioscorea alata*, 55
*Dioscorea bulbifera*, 152
Dioscoreaceae, 55, 152
*Diospyros revoluta*, 55, 76, 142, 148

Ebenaceae, 55, 142, 148
*Eleocharis mutata*, 147
*Elephantopus mollis*, 155
*Eleusine indica*, 110, 145
*Emilia coccinea*, 38, 153
*Emilia sonchifolia*, 38, 142, 154
*Erigeron canadensis*, 38, 143
*Ervatamia cumingiana*, 25, 144
*Eryngium foetidum*, 24, 117, 136
*Erythrina coralodendron*, 68
Erythroxylaceae, 154
*Erythroxylum havanense*, 154
*Eucalyptus resinifera*, 105, 141
*Eugenia cumini*, 106
*Eugenia jambos*, 106
*Eugenia malaccensis*, 107
*Eugenia sessiliflora*, 148
*Eugenia uniflora*, 105
*Eupatorium odoratum*, 38, 155
*Eupatorium villosum*, 38, 146, 152
Euphorbiaceae, 55, 137, 138, 140, 141, 142, 144, 145, 146, 147, 148, 149, 150, 154, 155, 156

*Euphorbia heterophylla*, 56, 142
*Euphorbia maculata var. thymifolia*, 56, 146
*Euphorbia oerstediana*, 148
*Euphorbia pulcherrima*, 57
*Euphorbia tirucalli*, 57, 144
*Euterpe globosa*, 9, 36, 143, 153
*Exostema elliptica*, 152

Fabaceae, 63, 135, 136, 137, 138, 139, 141, 142, 143, 144, 145, 146, 148, 149, 150, 151, 152, 153, 154, 155, 156
*Ficus aurea*, 103, 148
*Ficus nymphaeifolia*, 150
*Ficus stahlii*, 153
*Ficus trigonata*, 9
Flacourtiaceae, 89, 146, 149
*Flacourtia jangomas*, 89, 146, 149
*Flemingia strobilifera*, 155
*Fleurya aestuans*, 129, 145
*Fortunella margarita*, 119
*Freziera undulata*, 10

*Galactia filiformis*, 68, 145, 154
*Genipa americana*, 12, 117
Gentianaceae, 147
Gesneriaceae, 151
*Gleichenia dichotoma*, 155
*Gliricidia sepium*, 85, 145, 152

## Index of Scientific Names

*Gossypium barbadense*, 96, 97
*Guaiacum officinale*, 131, 149
*Guatteria caribaea*, 9, 151
*Guettarda elliptica*, 149

*Habenaria alata*, 156
*Haematoxylon campechianum*, 85, 149
*Hedyosmum arborescens*, 10
*Helianthus hirsutus*, 39, 140, 152
*Heliconia caribaea*, 81, 104, 147
*Heliconia subulata*, 148
*Heteropogon contortus*, 136
*Hibiscus rosa-sinensis*, 79, 96, 152
*Hibiscus sabdariffa*, 80, 97, 142, 148, 155
*Hibiscus tulipiflorus*, 9
*Hibiscus vitifolius*, 97, 150
*Hippobroma longiflora*, 49, 137
*Hippomane mancinella*, 57, 143
*Hura crepitans*, 9, 57, 77, 155
*Hylocereus trigonus*, 46, 146
*Hymenaea courbaril*, 9, 85, 141, 151

*Ilex opaca*, 18
*Ilex sideroxyloides*, 10
*Impatiens balsamina*, 40
*Indigofera suffruticosa*, 11, 85, 137
*Inga laurina*, 9, 86, 142, 151
*Ipomoea batatas*, 12, 52
*Ipomoea pes-caprae* subsp. *brasiliensis*, 53, 75, 135

*Ixora ferrea*, 9, 118

*Jacquinia armillaris*, 128
*Jacquinia berterii*, 128
*Jasminum multiflorum*, 107, 152
*Jatropha curcas*, 58, 138
*Jatropha gossypifolia*, 58, 138, 149
*Justicia pectoralis*, 15, 154

*Kalanchoe pinnatum*, 53

Lamiaceae, 89, 137, 139, 143, 145, 150, 152
*Lantana camara*, 129, 140, 154
*Lantana reticulata*, 129, 144
Lauraceae, 139, 141, 145, 149
*Leonotis nepetifolia*, 89, 143
*Licaria triandra*, 92, 142
Liliaceae, 93, 144, 146, 151, 153
*Lippia micromera*, 130
*Lippia nodiflora*, 130, 151
Loganiaceae, 95
Loranthaceae, 140
*Luffa aegyptiaca*, 54
*Lycopersicon esculentum*, 125
Lycopodiaceae, 145
*Lycopodium clavatum*, 145

*Macadamia ternifolia*, 115, 149
*Macfadyena unguis-cati*, 41, 144, 154

*Majorana hortensis*, 90
Malpighiaceae, 95, 139, 144, 149
*Malpighia punicifolia*, 96, 149
Malvaceae, 96, 136, 138, 142, 144, 146, 148, 150, 152, 155
*Malvastrum americanum*, 150
*Mammea americana*, 12, 51, 74, 136, 140, 151
*Mangifera indica*, 18, 19, 140, 142, 147, 150, 151
*Manihot esculenta*, 58, 59, 140
*Manilkara zapota*, 122
*Maranta arundinacea*, 98
Marantaceae, 98
*Marila racemosa*, 9
*Matelea maritima*, 37, 137
Melastomaceae, 136, 150, 155
*Melia azadirachta*, 99
*Melia azedarach*, 100, 138, 141, 142,
Meliaceae, 98, 138, 139, 140, 141, 142, 146, 153
*Melicoccus bijugatus*, 122, 142, 143
*Melocactus intortus*, 9, 47
*Mentha piperita*, 89, 139
*Mentha viridis*, 90
*Merremia dissecta*, 137
*Merremia umbellata*, 53, 139
*Miconia prasina*, 98, 150, 155
*Micropholis chrysophylloides*, 9
*Mimosa pudica*, 86, 138, 141

## Index of Scientific Names

*Mimusops coriacea*, 84, 122, 150
*Momordica charantia*, 54, 75, 123, 135, 152, 154
*Monstera adansonii*, 29, 146, 147
*Monstera deliciosa*, 29
*Monstera pertusa*, 29
Moraceae, 100, 140, 142, 147, 148, 149, 150, 153
*Morinda citrifolia*, 118, 150, 152
Moringaceae, 103, 148
*Moringa oleifera*, 103, 148
*Morisonia americana*, 153
*Mucuna pruriens*, 86
Musaceae, 104, 147, 148
*Musa paradisica*, 104
*Musa sapientum*, 91, 104, 105
*Myrcia splendens*, 105, 142
Myrtaceae, 105, 139, 141, 142, 148, 150, 151, 152, 155

*Nectandra coriacea*, 92, 145
*Nectandra membranacea*, 93, 139
*Nepeta cataria*, 90, 137, 139, 152
*Nerium oleander*, 25, 148
*Nicotiana tabacum*, 126, 153

Ochnaceae, 155
*Ocimum micranthum*, 90
Oleaceae, 152
*Oncidium altissimum*, 141

*Opuntia*, 9
*Opuntia cochenillifera*, 47, 48
*Opuntia dillenii*, 48
Orchidaceae, 141, 147, 156
*Origanum majorana*, 90, 139
*Origanum vulgare*, 91, 139, 145

*Palicourea crocea*, 10, 136, 155
*Parthenium hysterophorus*, 39, 152
Passifloraceae, 108, 138, 141, 143, 152
*Passiflora edulis*, 108, 143
*Passiflora foetida*, 108, 138, 141, 152
*Peperomia pellucida*, 109, 136, 140
*Persea americana*, 12, 93
*Petiveria alliacea*, 109
*Phaseolus lunatus*, 12, 86
*Phaseolus vulgaris*, 12, 86
*Philodendron giganteum*, 29, 70
*Phoenix dactylifera*, 36, 37
*Phoradendron trinervium*, 45, 95, 140
*Phyllanthus acidus*, 62, 149
*Phyllanthus anderssonii*, 62, 145
*Phyllanthus mimosoides*, 9, 156
Phytolaccaceae, 109
*Picrasma antillana*, 123, 151
*Pimenta racemosa*, 9, 105, 152, 155
*Pimpinella anisum*, 24
Piperaceae, 109, 136, 140
*Piper amalago*, 136
*Piper dilatatum*, 109, 140

*Piscidia carthagenesis*, 87, 153
*Piscidia piscipula*, 78, 87, 138, 146, 153
*Pitcairnia angustifolia*, 10, 150
*Pithecellobium arboreum*, 87
*Pithecellobium dulce*, 88
*Pithecellobium saman*, 88, 138, 141, 155
*Pithecellobium unguis-cati*, 88, 149
Plantaginaceae, 110
*Plantago major*, 110
*Pluchea carolinensis*, 39
*Plumeria alba*, 9, 25, 144
*Plumeria bahamensis*, 146
*Plumeria rubra*, 26, 69, 144
Poaceae, 110, 136, 137, 138, 139, 141, 142, 144, 145, 147, 148, 152
*Podocarpus coriaceus*, 9
*Pogonia rosea*, 146
Polygonaceae, 114, 135, 151, 156
Polypodiaceae, 114, 140, 147, 155
*Polypodium heterophyllum*, 114, 140
Portulacaceae, 115, 139, 141, 156
*Portulaca oleracea*, 115, 139, 141
*Prosopis juliflora*, 88
Proteaceae, 115, 149
*Pseudelephantopus spicatus*, 39, 154
*Psidium guajava*, 12, 106, 139, 150, 151
Psilotaceae, 115, 147
*Psilotum nudum*, 115, 147
*Psychotria domingensis*, 155

## Index of Scientific Names

*Pterocarpus indicus*, 88, 149
Punicaceae, 116, 148
*Punica granatum*, 116, 148

*Randia aculeata*, 118, 145
*Rauvolfia nitida*, 26
Rhamnaceae, 116
*Rhyticocos amara*, 9
*Richeria grandis*, 9
*Ricinus communis*, 62, 140
Rosaceae, 116, 145
*Rosmarinus officinalis*, 91, 139
Rubiaceae, 117, 136, 137, 140, 145, 147, 149, 150, 151, 152, 155, 156
*Rubus rosifolius*, 116, 145
*Ruellia tuberosa*, 15, 143, 153, 154
Rutaceae, 118, 140, 141, 148

*Saccharum officinarum*, 82, 111, 142
*Sambucus simpsonii*, 50, 143
*Sansevieria metallica*, 95, 144
Sapindaceae, 120, 142, 143, 146
*Sapium caribaeum*, 63
Sapotaceae, 122, 150
*Sauvagesia erecta*, 155
Scrophulariaceae, 122, 142
*Sechium edule*, 55, 76, 143
Selaginellaceae, 147
*Selaginella eatoni*, 147

*Selaginella substipitata*, 147
*Sida ciliaris*, 98, 137, 142, 144
*Simarouba amara*, 9, 123, 144, 145
Simaroubaceae, 123, 144, 145, 152
*Sloanea*, 9
*Sloanea dentata*, 9
Smilacaceae, 123, 150
*Smilax cumanensis*, 29, 123, 150
Solanaceae, 124, 136, 138, 141, 143, 146, 150, 153
*Solanum ciliatum*, 119, 126
*Solanum erianthum*, 150
*Solanum ficifolium*, 126, 136
*Solanum melongena*, 126, 127
*Solanum nigrum*, 143
*Solanum nodiflorum*, 143
*Solanum tuberosum*, 12
*Spigelia anthelmia*, 95
*Spondias cytherea*, 19, 149
*Spondias mombin*, 12, 20, 149
*Spondias purpurea*, 20, 139
*Stachytarpheta cayennensis*, 130, 136, 143, 145
*Stachytarpheta jamaicensis*, 130, 151
Sterculiaceae, 126, 137, 151
*Swietenia macrophylla*, 100, 146
*Swietenia mahagoni*, 9, 153
*Syzygium cumini*, 106
*Syzygium jambos*, 106, 139

*Syzygium malaccense*, 107, 148

*Tabebuia pallida*, 9, 41, 146
*Talinum triangulare*, 156
*Tamarindus indica*, 89, 138, 154
*Tanacetum vulgare*, 39, 139
*Tecoma stans*, 41, 156
*Tectonia grandis*, 130
*Terminalia catappa*, 52, 152
*Ternstroemia peduncularis*, 10
*Theobroma cacao*, 126, 127, 128, 137, 151
Theophrastaceae, 128
*Thespesia populnea*, 98
*Thevetia peruviana*, 26, 144
*Thunbergia alata*, 135, 155
*Thymus vulgaris*, 91, 150
*Torulinium odoratum*, 136
*Tournefortia filiflora*, 43, 145
*Tournefortia volubilis*, 44
*Toxicodendron radicans*, 18
*Tragia volubilis*, 63, 141
*Trichachne insularis*, 113, 136, 138, 142, 144, 147
*Triphasia trifolia*, 120, 141
Tropaeolaceae, 129, 153
*Tropaeolum tuberosum*, 129, 153
*Turbina corymbosa*, 53, 146

Urticaceae, 129, 145

## Index of Scientific Names

Verbenaceae, 129, 136, 140, 143, 144, 145, 151, 153, 154
*Vernonia cinerea*, 40, 156
*Vetiveria zizanioides*, 113, 141
Vitaceae, 131, 137, 146, 156
*Vittaria lineata*, 147
*Voyria tenella*, 147

*Wedelia trilobata*, 40, 145, 154

*Xanthosoma brasiliense*, 29, 148
*Xanthosoma sagittifolium*, 30

*Zanthoxylum flavum*, 120
*Zanthoxylum monophyllum*, 120, 140
*Zea mays*, 12, 113
Zingiberaceae, 131, 144, 148
*Zingiber officinale*, 131, 148
Zygophyllaceae, 131, 149

**David Eric Brussell** holds a master's degree in botany from Eastern Illinois University and a Ph.D. in botany from Southern Illinois University at Carbondale. He is a visiting assistant professor of plant biology at Southern Illinois University at Carbondale in Niigata, Japan, and an adjunct assistant professor of plant biology at Southern Illinois University at Carbondale.

Brussell has taught at colleges and universities in Illinois, Nebraska, Virginia, Japan, and Mexico. He was a senior biologist (lecturer in ethnobiology, taxonomy, and natural history) at Pacific Quest Inc., Haleiwa, Hawaii. He was chosen as a Fulbright Senior Research Scholar to conduct a study of medicinal plants in Greece. He directed the natural history program and managed the nature preserve when he served as the naturalist for the Mitchell Foundation and Museum. Archaeological projects he has worked on include Black Mesa in Arizona and Carrier Mills in Illinois.

Brussell has published articles in both professional journals and popular periodicals and has contributed photographic work to *Natural History* magazine. His current research interests include ethnobotanical investigations in the Caribbean, Japan, Greece, Turkey, Egypt, Polynesia, Latin America, and the Swiss Alps. In addition, he is involved in teaching summer travel study courses in Greece, Egypt, French Polynesia, Spain, Portugal, Morocco, Hawaii, and Turkey on the history of medicine, plant classification, and ethnobiology.

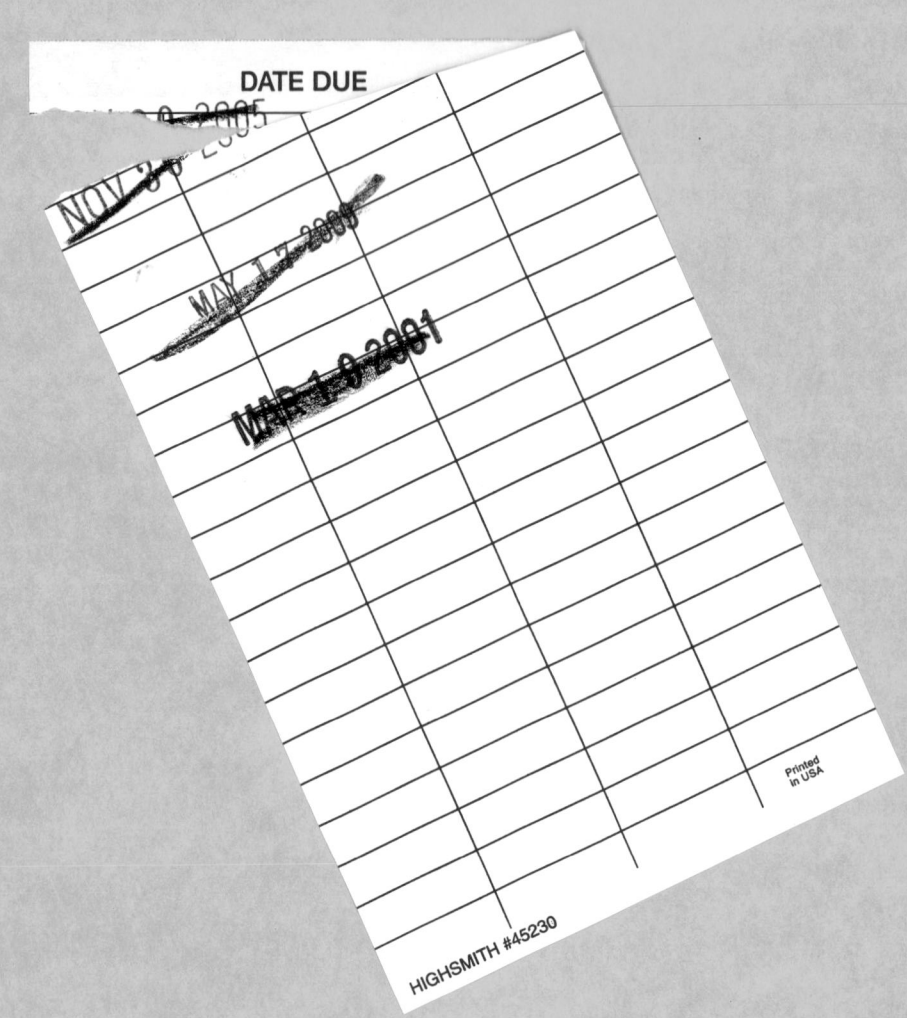